A Visual Atlas for Soil Micromorphologists

Eric P. Verrecchia • Luca Trombino

A Visual Atlas for Soil Micromorphologists

 Springer

Eric P. Verrecchia
Institute of Earth Surface Processes
University of Lausanne
Lausanne, Switzerland

Luca Trombino
Department of Earth Sciences
University of Milan
Milan, Italy

ISBN 978-3-030-67805-0   ISBN 978-3-030-67806-7   (eBook)
https://doi.org/10.1007/978-3-030-67806-7

This Springer imprint is published by the registered company Springer Nature Switzerland AG
The registered company address is: Gewerbestrasse 11, 6330 Cham, Switzerland

For Milena

Foreword

Micromorphology, the microscopic investigation of undisturbed earth materials, is by definition based on the ability to identify components and to recognize shapes, arrangements, and patterns in thin sections. Microscopic observation is complicated by the fact that a two-dimensional image is used to observe a three-dimensional reality. A book with reference images can, therefore, be of invaluable importance for micromorphologists.

In the past, handbooks on micromorphology were sparsely illustrated with black and white photographs. It is only since the beginning of this century that the use of colour plates became economically feasible. Although some initiatives were taken to make more reference images available for students and researchers, they only reached a limited audience.

In life sciences, such as medicine, biology, botany, and wood anatomy, atlases of microscopic images have existed since the early twentieth century, the earliest of which often included coloured drawings. Similarly for mineralogy and petrography, atlases of rocks and mineral images under the microscope were published in the second half of last century and were used with enthusiasm by generations of students. Such an atlas is missing for soil micromorphology. The initiative taken by Eric Verrecchia and Luca Trombino is, therefore, more than welcome. This atlas has been prepared not only for beginner soil micromorphologists but also for more experienced researchers. Images are complemented by informative text explaining concepts and terms, and by references to the literature, and where necessary, a historic insight into the evolution of the terminology. A list of translations of the terms into French, Italian, and German at the end of the book will contribute to widen its use internationally.

Ghent, Belgium   Prof. Em. Georges Stoops

Acknowledgements

Many people provided samples or thin sections to complement our own collection, which were indispensable to be able to illustrate the large variety of features observed in thin sections of soil: Yann Biedermann (UniNe[1]), Dr. Filippo Brandolini (UniMi[2]), Dr. Guillaume Cailleau (DataPartner, CH), Prof. Mauro Cremaschi (UniMi), Dr. Nathalie Diaz (Unil[3]), Dr. Fabienne Dietrich (Unil), Prof. Alain Durand (Université de Rouen, F), Dr. Laurent Emmanuel (Sorbonne Université, F), Dr. Stephania Ern (Cantone Ticino, CH), Dr. Katia Ferro (UniNe), Prof. Karl Föllmi (Unil), Prof. Pierre Freytet[4] (Université Paris-Sud Orsay, F), Prof. Jean-Michel Gobat (UniNe), Dr. Stephanie Grand (Unil), Céline Heimo (UniNe), Dr. Guido Mariani (UniMi), Dr. Loraine Martignier (Unil), Dr. Anna Masseroli (UniMi), Dr. Ivano Rellini (Università degli Studi di Genova, I), Rémy Romanens (Unil), Dr. David Sebag (Université de Rouen, F and Unil), Dr. Brigitte Van Vliet-Lanoë (CNRS, Université de Bretagne Occidentale, F), Prof. Andrea Zerboni (UniMi), and Dr. Luisa Zuccoli Bini (MIUR, I).

Soil micromorphology will continue to need the talent of gifted technicians, engineers, and researchers. We would like to thank our colleagues who provided documents or spent time with us on specific techniques: Dr. Benita Putlitz (Unil), Dr. Daniel Grolimund (PSI, CH), Dr. Kalin Kouzmanov (Université de Genève, CH), Dr. Laurent Remusat (Muséum National d'Histoire Naturelle, F), Dr. Alexey Ulyanov (Unil), and Dr. Pierre Vanlonthen (Unil). We would like to thank the students of the MSc in Biogeosciences program (Universities of Lausanne and Neuchâtel) who kindly chose the title of this Atlas and tested its draft version, Titi, Scintillina, and the Dragon for their valued support.

The authors benefited from funding through different sources during the making of this Atlas, which has been written in Lausanne within the framework of a scientific agreement between the universities of Lausanne and Milan (special thanks to Denis Dafflon and Marc Pilloud, International Relations, and Prof. François Bussy, the Faculty of Geosciences and the Environment, all from the University of Lausanne). The *Fondation Herbette* funded stays for Prof. Luca Trombino in Lausanne. The Swiss National Science Foundation made possible free access for the e-version of the Atlas by funding a Gold Open Access agreement with Springer-Nature. Special thanks to Zachary Romano (Springer-Nature), who believed in our project, supported us, and edited our Atlas. His help and his kindness made this adventure much easier. Finally, we would like to thank Karin Verrecchia for her endless patience and her careful proofreading of the manuscript.

If Prof. Georges Stoops had not been such a great scientist, a wonderful teacher, and such an endearing person, the authors would have never met and probably not considered soil micromorphology to be as important and relevant as it really is. Thank you Georges for your endless help and consideration.

---

[1] UniNe stands for Université de Neuchâtel, Switzerland.

[2] UniMi stands for Università degli Studi di Milano, Italy.

[3] Unil stands for Université de Lausanne, Switzerland.

[4] Profs Karl Föllmi and Pierre Freytet sadly passed away shortly before the publication of this Atlas.

Introduction to the Atlas

Why Use Such an Atlas?

Natural sciences are based on the observation of natural objects. The precise description of their characteristics is fundamental in order to establish nomenclatures. From these nomenclatures, the study of the processes at the origin of their distinctive features allows classifications: classifications are built using qualitative, quantitative, and semi-quantitative parameters of specific features, which allow hierarchical relationships between objects to be drawn. Consequently, before pretending to understand the origin of a natural object, it is necessary to identify its borders, describe its properties, and compare it to other similar objects belonging to the same nomenclature. Soils are no exception. Unfortunately, many soil scientists contend that going directly from the hand lens observation in the field to the mass spectrometer analyses in the lab fills all the requirements for a suitable and thorough investigation. They are wrong.

Indeed, soils constitute a unique and emergent property of the complex interactions between life and mineral matter. Only looking at soils from the inside, in their minute detail and at various microscopic scales, allows soils to be explored with the best acuity. A simple example: measuring the amount of calcium carbonate in a soil does not say anything about the location and origin of this calcium carbonate. Is it along the pores, as tiny nodules or in the groundmass as impregnations? Is it micrite or needle-fibre calcite associated with fungi, a sparitic coating or calcified root cells? All this information is not available if the investigator cannot observe the structure of the objects themselves, using the appropriate tool. Crushing and grinding a soil sample to a very fine powder provides information about its chemistry and the nature of some of its compounds but reveals nothing about the relationships, the organization, and the hierarchy of the various features and objects that constitute its architecture and record its history.

Moreover, according to Richter and Yaalon (2012), soils are all polygenic paleosol systems, superimposed over time, forming a sort of palimpsest. Therefore, there are traces of old mechanisms, like a permanent background noise, which alters the geochemical signal of the contemporary dynamics. Consequently, the question must be asked: how much importance should be given to "blind" (i.e. bulk) geochemical studies that consider the soil as a functional, single-phase continuum? What is the meaning of using, for example, the τ factor (Brantley et al., 2007), when the parent material remains as a trace component or a phase impossible to clearly identify and when the bulk fraction results from a diachronic mixture? A better method would be to consider the use of soil micromorphology, which allows the soil to be seen from the inside and to identify the traces of past pedogenesis. Such an approach would allow the geochemical analyses to target objects indicative of such past pedogeneses. This method requires an extensive experience to address the qualitative issues related to the selection of the pertinent and most promising pedofeatures. It justifies further access to often expensive equipment (micro-drill sampling, microprobe and synchrotron investigations, mass spectrometry on very small quantities, laser-ablation ICP-MS on thin sections, etc.), in order to quantitatively characterize the elementary dynamics at work in the selected pedofeatures and recombinations of trace quantities. In conclusion, soil micromorphology affords most of the necessary tools, vocabulary, and methods of observation that will facilitate the investigations. This practical Atlas aims at providing the necessary comparative and visual references to guide the soil micromorphologist in her or his identification of the various soil objects observed under the microscope. It does not aim at providing interpretations. Instead, it proposes to relate concepts and vocabulary of soil micromorphology to images of the real soil world. Therefore, the Atlas helps the micromorphologist to apply concepts and vocabulary in a rigorous manner by using comparisons between her or his own thin sections with a collection of examples. Nonetheless, Stoops et al. (2018) presented a comprehensive reference for interpretations, once features have been properly described and identified. This Atlas is, therefore, complementary and must be used before opening Stoops et al. (2018).

This Atlas is designed for researchers, academics, and students at the master's and doctoral levels, so they can rapidly find features and structures observed in thin sections of soil. It is convenient for fast self-instruction by using comparative photographs. Therefore, it can also be used in the classroom as a visual resource book, the eye being the best tool for learning natural features by intuitive links of shapes and colours, or as a reference for comparisons in advanced studies. Therefore, this Atlas provides a basic background to build a pertinent nomenclature, which will help to identify the process-oriented challenges associated with soils. Finally, the reader must keep in mind that soil micromorphology is more than a scientific method to investigate soils. It is also a way of envisaging natural sciences. The method itself requires time, in contrast to a lot of today's "fast science". The soil micromorphologist has to wait for the thin section fabrication and then has to spend hours with the microscope, acquiring the experience necessary to identify the myriad features that appear in nature. This is the profession of the *The Slow Professor*.[5]

Online Database and Digital Resources in Soil Micromorphology

Although many websites are available for images of rock-forming minerals under the microscope, there are only a few dealing with soil micromorphology, e.g. edafologia.ugr.es/english/index.htm or spartan.ac.brocku.ca/~jmenzies. Moreover, there are many websites describing and explaining the principles of optical microscopy: the following webpage of the Soil Science Society of America proposes a large choice of such websites: www.soils.org/membership/divisions/soil-mineralogy/micromorphology. Georges Stoops' handbook, in its first edition (Stoops 2003), was accompanied by a CD-ROM with many micromorphological images. Unfortunately, today, most computers do not include CD-ROM readers anymore, so it seemed necessary to provide soil micromorphologists with an atlas in the form of a printed book and/or an e-book with high-resolution images. Indeed, this Atlas is available as an *Open Access* pdf section at the Springer-Nature website: the high-resolution images provide details at high magnification making the e-book easy to use during observations on a tablet computer.

Today, access to powerful computers makes possible the use of image analysis to quantify features and textures. Most of these software are presently proposed as multiplatform applications. Over the last few years, ImageJ (http://imagej.nih.gov/ij/download.html), or its bundled version Fiji (https://imagej.net/Fiji), became one of the most used freeware in image analysis. It replaces NIH-Image, its ancestor, but some of the macros can still be run on the appropriate version of computers (Heilbronner and Barrett 2014). Gwyddion (http://gwyddion.net/) is another freeware that can be used in image analysis. For people who like to generate code, Scilab remains an extremely interesting open-source solution (http://www.scilab.org/) and can advantageously replace the powerful and user-friendly, but costly, Matlab®. Of course, there are multiple commercial software, some of them being sometimes fairly expensive and provided as a closed system. Therefore, this choice is not necessarily the most appropriate for teaching and research in the academic environment.

---

[5] Berg M. and Seeber B.K. (2016) *The Slow Professor—Challenging the Culture of Speed in the Academy*. University of Toronto Press. Toronto, Canada.

How to Use This Visual Atlas

Terminology Used in the Atlas

The micromorphological terminology used in this Atlas is mostly based on Stoops (2003, 2021). Nevertheless, some concepts or keywords also refer to Bullock et al. (1985) and Brewer (1964), as they provide complementary vocabulary and a different kind of logic applied to the description. Older textbooks contain the descriptions on which most of the present-day soil micromorphology was built. They are just as pertinent today and should not be overlooked.

Book Structure

The Atlas is organized into six chapters (including Annexes), and each chapter is divided into sections. Each section contains a series of images, usually eight, on the left-hand page, and an explanatory text on the right-hand page. Regarding the microphotographs, they are usually displayed in plane-polarized light (PPL) and cross-polarized light (XPL), if not specified otherwise. PPL and XPL views are usually presented as two halves of the same microphotograph, separated along the diagonal. The upper half is always the PPL view and the right lower, the XPL one. Moreover, microphotographs are shown as observed under the microscope, without any alteration, such as arrows, letters, or numbers. The choice of pristine images, such as in MacKenzie et al. (2017), has been made in order to provide self-explanatory views. The text on the right-hand page supplies all the needed information and/or explanation. In addition, each chapter is introduced by a short paragraph in a grey box summarizing the main concepts. All the microphotographs, if not mentioned otherwise, have been taken with an Olympus BX53 polarizing microscope or an Olympus stereomicroscope SZX16 system, both equipped with an Olympus DP73 digital camera operated by Olympus cellSens imaging software.

The six different chapters of the book are devoted to different aspects of the micromorphological approach to studying soils. The technical aspects are presented in Chap. 1: they consist of the sampling strategy for soil profiles, the preparation of thin sections, the various tools used in optical microscopy, and finally the micromorphological approach, which is detailed in a flow chart. The second chapter is related to the organization of soil material, i.e. the fabric, the c/f related distribution, aggregates, voids, and microstructures. In Chap. 3, both mineral and organic constituents are presented in terms of size, sorting, and shape. In addition, this chapter introduces their various natures, whether they are rocks, mineral micromass and grains, biominerals, anthropogenic features, or organic matter. The fourth chapter is a list of pedogenic features as imprints of pedogenesis, presented according to their nature and morphology, e.g. clay coatings, biogenic infillings, or iron nodules. The fifth chapter provides some examples of features associated to the main soil processes observed in thin sections: the imprint of water, the influence of clays, the precipitation of carbonate, gypsum, and oxyhydroxides, and biogeochemical processes. The short Chap. 6 presents a view of what the future of soil micromorphology could be when thin sections are used with instruments other than the conventional optical microscope, such as electron microprobes or laser-ablation ICP-MS. Finally, the Annexes list the formula of the main soil minerals, present some common errors and pitfalls, and propose a way to describe thin sections accurately. A four-language list of micromorphological terms, which can be used to facilitate translations, is found at the end.

References

Brantley, S., Goldhaber, M., & Ragnarsdottir, K. (2007). Crossing disciplines and scales to understand the critical zone. *Elements, 3*, 307–314.

Brewer, R. (1964). *Fabric and mineral analysis of soils*. London: John Wiley and Sons.

Bullock, P., Fedoroff, N., Jongerius, A., Stoops, G., & Tursina, T. (1985). *Handbook for soil thin section description*. Wolverhampton: Waine Research Publications.

Heilbronner, R., & Barrett, S. (2014). *Image analysis in earth sciences*. Berlin: Springer-Verlag.

MacKenzie, W., Adams, A., & Brodie, K. (2017). *Rocks and minerals in thin section: a colour atlas*. Boca Raton: CRC Press.

Richter, d. D., & Yaalon, D. (2012). The changing model of soil, revisited. *Soil Science Society of America Journal, 76*(3), 766–778.

Stoops, G. (2003). *Guidelines for analysis and description of soil and regolith thin sections*. Madison: Soil Science Society of America, Inc.

Stoops, G. (2021). *Guidelines for analysis and description of soil and regolith thin sections*. Second Edition, Wiley.

Stoops, G., Marcelino, V., & Mees, F. (Eds.) (2018). *Interpretation of micromorphological features of soils and regoliths* (2nd ed.) Amsterdam, NL: Elsevier.

Contents

About the Authors

Eric P. Verrecchia is a full professor of Biogeosciences at the Faculty of Geosciences and the Environment, University of Lausanne (Switzerland). He is specialized in geopedology and biogeochemistry of the terrestrial carbon and calcium cycles. Awarded by a Marie-Curie Fellowship for Senior Researchers in 1994–1995, he joined Prof. G. Stoops' laboratory of soil micromorphology at the Ghent University (Belgium), where he was introduced to this microscopic approach to study soils. Since then, he applied this technique, coupling it with biogeochemical methods, to soils from the tropics to the temperate zone, particularly in calcium- and carbonate-rich environments.

Luca Trombino is professor in the Earth Sciences Department at the Universita degli Studi di Milano (Italy). His main research topics are in the field of paleopedology, soils, and archaeological deposits, where he extensively uses thin section micromorphology, coupled to sedimentology and Quaternary geology methods. He started to practice soil micromorphology in 1994 when he attended the courses by Prof. G. Stoops at the Ghent University (Belgium) and, at present, is teaching Micropedology to the students of M.Sc. in BioGeoSciences and M.Sc. in Conservation Science at the University of Milan.

Chapter 1

Observation of Soils: From the Field to the Microscope

File 1: The Multiscalar Nature of Soils

As emphasized by W. Kubiëna, "... there exists no other method capable of revealing the nature and complexity of soil polygenesis in so much detail as thin-section micromorphology and at the same time enabling one to follow and explain its formation...". This sentence, cited by Fedoroff (1971), highlights the aim of soil micromorphology: looking at a soil from the inside and at various scales, from the optical microscope to synchrotron imaging. Soils constitute multiscalar objects by definition, from their soilscape (at the landscape scale), to their profile and its horizons to the atomic interactions between the smallest minerals and organic molecules. Micromorphology enters the soil investigations at the multi-centimetre scale (see "File 3") at which the thin section is made. The scales of observation span from the millimetre to the micrometre, and even down to the nanometre using electron microscopy (see "File 7" and "File 8"). Therefore, the micromorphological approach is based on multiscalar observations because the different features and properties of soils require different magnifications; in addition, this approach is twofold using composition and fabric (see "File 9").

Early microscope

Cartoon from the Larsen's Far Side. This cartoon has been published in the Proceedings of the International Working Meeting on Soil Micromorphology in San Antonio, Texas, July 1988.

E. P. Verrecchia, L. Trombino, *A Visual Atlas for Soil Micromorphologists*,
https://doi.org/10.1007/978-3-030-67806-7_1

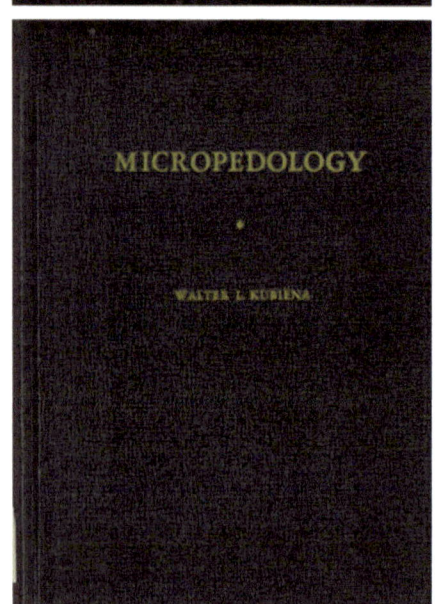

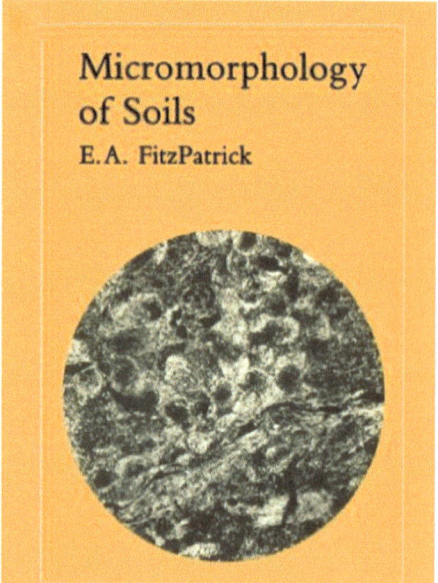

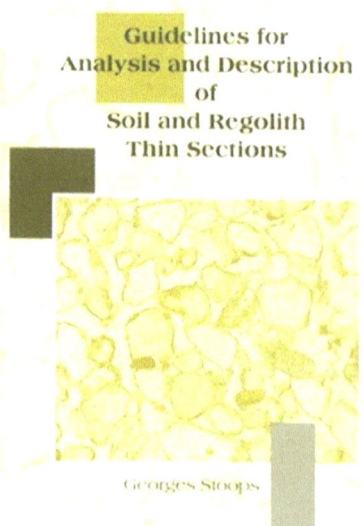
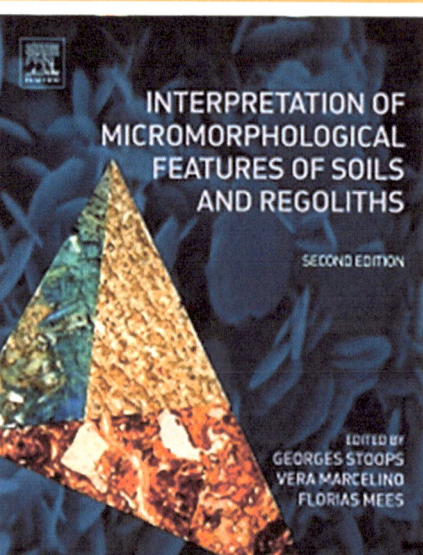

File 2: History of Micromorphology

Soil micromorphology is a relatively recent method, first popularized by a book by Kubiëna (1938). This method revolutionized the way that a soil was observed because it was studied from the inside, i.e. the inner organization. Indeed, during the "Symposium on the Age of Parent Material and Soils", in 1970, Walter Kubiena, the father of the micropedology, stated: "... there exists no other method capable of revealing the nature and complexity of soil polygenesis in so much detail as thin-section micromorphology and at the same time enabling one to follow and explain its formation ..." (Fedoroff 1971).

Captions from upper left corner to lower right corner.

1. Photograph of Walter L.K.R. von Kubiëna (1897–1970): considered as the father of soil micromorphology, W. Kubiëna investigated microscopic crystal formation and neoformation in soils during his several stays in the USA (mostly during 1931 and 1933). The method he applied is based on thin sections and microscopic observation with a polarizing microscope, which provided him with new insights into the variety of forms encountered in soils. During the next few decades, he made soil micromorphology his main field of research. Invited to give lectures at Iowa State College in Ames (USA), he decided to gather his experience into a book, *Micropedology* (Kubiëna 1938), which became the first international standard work in soil micromorphology. (Photo credit: www.spektrum.de)
2. Photograph of Ewart A. FitzPatrick (1926–2018): author of one of the most successful books in soil micromorphology published in 1984, and simply titled *Micromorphology of Soils* (FitzPatrick 1984), the work by E. A. FitzPatrick is renown for its clarity, magnificent illustrations, and its simple and user-oriented approach, avoiding complex vocabulary. E. A. FitzPatrick also developed techniques and protocols optimising the fabrication and interpretations of thin sections. He received the Kubiëna Medal of the International Union of Soil Sciences in 1996. (Photo credit: International Union of Soil Sciences)
3. Photograph of Georges Stoops: as a major promoter of soil micromorphology during the last 40 years, G. Stoops fundamentally contributed to the standardization of the vocabulary used in thin section description, as well as to the classification and interpretation of micromorphological features (Stoops 2003). He received the Kubiëna Medal of the International Union of Soil Sciences in 1992.
4.–9. Covers of the main milestone books in soil micromorphology in a chronological order: Kubiëna (1938), Brewer (1964), FitzPatrick (1984), Bullock et al. (1985), Stoops (2003), and Stoops et al. (2018).

File 3: Observation and Sampling of Soils

There are various ways to sample soil profiles in the field to get the right soil portion of interest that represents the horizon variability. Sampling can focus on horizons, transitions, or specific soil features. The most conventional sample selection is based on the use of Kubiëna boxes of various shapes, sizes, and compositions. They are mainly used in soils with fine textures. However, the soil texture can sometimes be an obstacle to conventional field sampling: a manual block extraction sampling technique is then performed in order to get enough soil material for further impregnation in the laboratory. Moreover, a plaster-impregnated patch of burlap can be used to encase and sample a large block of undisturbed soil (Goldberg and Macphail 2003).

Captions from upper left corner to lower right corner.

1. Profile of a podzol from Fontainebleau Forest (Paris Basin, France) with its six horizons (O, A, E, Bh, Bs, C). The E horizon's lower boundary has deep vertical tongues. Below the easily observed E and Bh horizons, there are bands enriched in iron, corresponding to the diffusion of organometallic complexes following grain-size irregularities of the parent material. Scale = 80 cm.
2. Examples of sampling locations on the profile described above. The three rectangles on the left side indicate the sampling location of each horizon. The two boxes in the middle show the horizon's transitions. Finally, the right side boxes point out special features, i.e. tongues and mottles of organometallic complexes.
3. Homemade Kubiëna boxes ready to collect undisturbed samples: these are empty solid square (8 × 8 cm) aluminium frames with sharp edges. Lids are used to preserve the undisturbed soil after sampling. An arrow must be drawn on the box pointing to the top (i.e. the ground's surface). A sample number can also be written directly on the box's lid with a felt pen. Such boxes can be easily cleaned with water and ethanol and reused.
4. Homemade Kubiëna boxes of different materials inserted directly in a soil profile. Left: metal boxes in a loamy soil; right: plastic boxes in a silty soil.
5. Left: insertion of a plastic Kubiëna box in a soil profile with the help of a rubber hammer. Right: extraction of the box with the help of a trowel (a knife can equally be used); after having clearly delimited the border of the box, it is gently extracted to avoid any disturbance.
6. Sampling of a large soil block without a field Kubiëna box: such samples must be wrapped with appropriate protection; back in the laboratory, they will be directly impregnated in a tray.
7. From left to right: example of the extraction of a block from a loamy soil profile with a trowel.

File 4: How to Make Thin Sections

> The micromorphological observation of soils starts with the fabrication of thin sections. This process is generally long because soil samples must be impregnated to solidify them: a polystyrene-based resin, often with a density close to water, is poured on samples. With time, it will penetrate inside the soil pores and harden. Making thin sections requires some specific equipment. The series of photographs included here provides only the major steps involved in thin section preparation. For more information see Murphy (1986) and Benyarku and Stoops (2005).

Captions from upper left corner to lower right corner.

1. Impregnation of soil blocks inside trays filled with polystyrene-based resin with an added accelerator and a catalyst (laboratory at the University of Ghent, Belgium). Acetone can also be used as a diluent, if needed.
2. Vacuum system used to remove residual air from impregnated samples at the University of Lausanne, Switzerland. This process is often used when the polystyrene-based resin is still liquid, as vacuum can help to access micropores and partly avoid problems related to air bubbles.
3. Impregnated and hard samples ready for sawing (laboratory at the University of Ghent, Belgium). The arrow shows the top of the soil.
4. Diamond saw machine used to prepare slabs for grinding, polishing, and lapping at the University of Lausanne, Switzerland. The aim is to obtain slabs slightly smaller than the size of the glass slide chosen to mount the thin section, and of a thickness depending on the type of saw, but generally about 1–2 cm.
5. Lapping machine used to flatten the slabs and frost the glass slide (laboratory at the University of Lausanne, Switzerland).
6. Polishing machine for ultra-smooth surfaces (laboratory at the University of Lausanne, Switzerland). For example, such thin sections are used for cathodoluminescence (see "File 6") or microprobe chemical imaging (see "File 7").
7. Examples of various sizes of soil thin sections. The most common sizes are the following: standard (28×48 mm), medium (60×90 mm), large (90×120 mm), mammoth (120×180 mm).
8. Example of the succession of phases needed to prepare soil thin sections: (1) slab extract from impregnated blocks using diamond saw machines; (2) first face lapping; (3) bonding of lapped face to frosted glass slide; (4) thinning of the bulk face; (5) grinding and lapping of the thin section; (6) thin section polishing; after this phase, the thin section is $30\,\mu m$ thick and ready for observation. Thin sections can be glass-covered or not. Uncovered thin sections are usually used for staining, selective extraction (see "File 8"), cathodoluminescence (see "File 6"), backscattered electron or chemical imaging (see "File 7", "File 77", and "File 78").

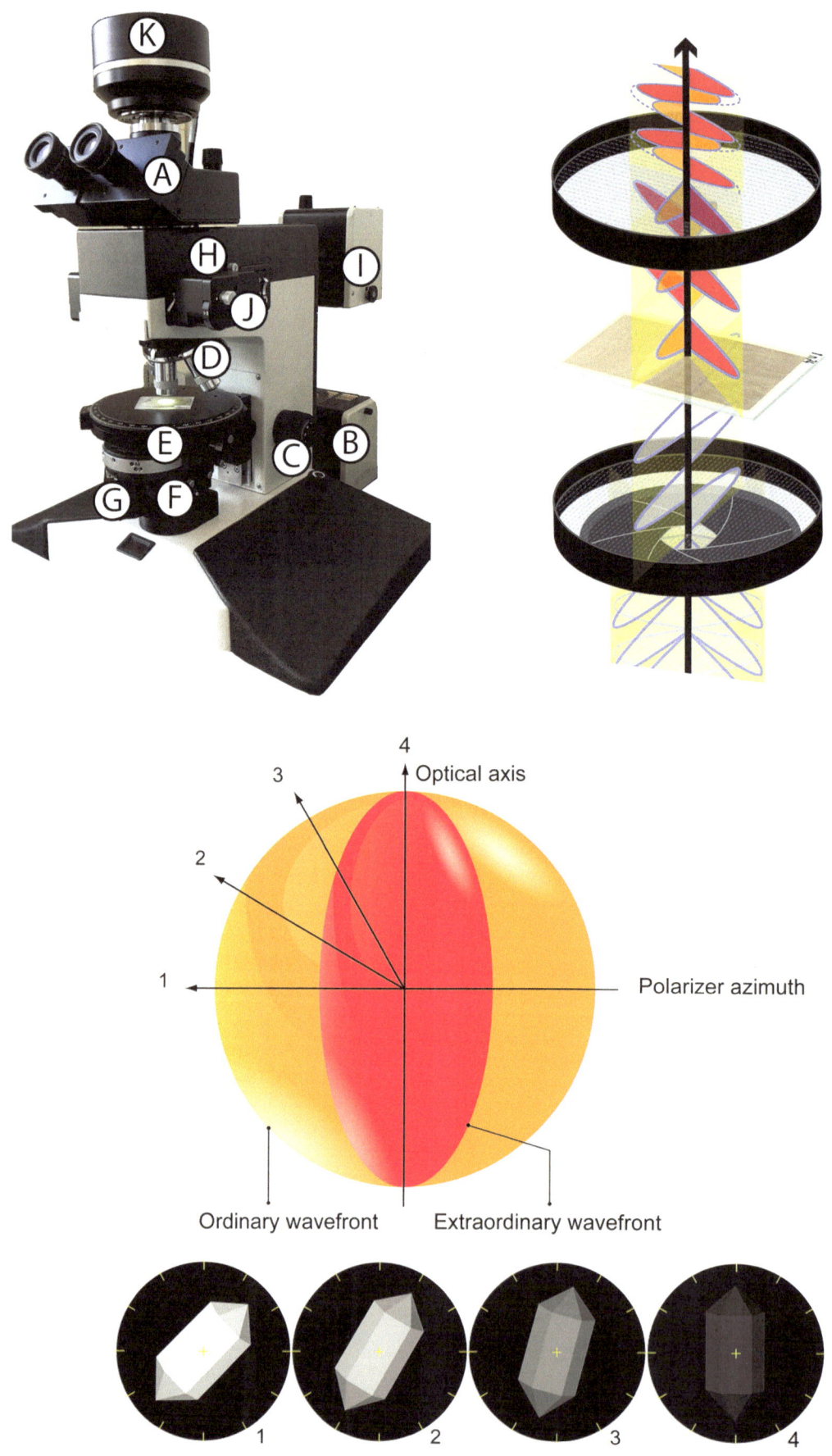

Optical axis

Polarizer azimuth

Ordinary wavefront   Extraordinary wavefront

File 5: The Polarized Light Microscope

In order to perform micromorphological observations, the polarized light microscope is the most appropriate tool. It couples the magnification of a conventional optical microscope with light polarization induced by a polarizer and an analyser located along the optical light pathway. Micromorphologists use the optical anisotropic properties of soil constituents for their identification and the observation of their potential transformation.

Captions from upper left corner to lower right corner.

1. An example of a fully equipped polarized light microscope with fluorescence: (A) binocular observation tubes (eyepieces), (B) source of transmitted light, (C) focus, (D) revolving objectives, (E) circular rotating stage, (F) 360° rotating polarizer, (G) condenser, (H) analyser slide, (I) source of incident light (used for fluorescence), (J) fluorescence filters, (K) microphoto or video camera.
2. Principle of light polarization with two polarizing filters: first, the light comes from a white source at the bottom of the sketch. There are two polarizing filters along the optical pathway, the polarizer and the analyser. The polarizer is below the slide stage and fixes the vibration direction of light in a single direction, let us say north–south. The analyser allows vibration of the light perpendicular to the polarizer direction (i.e. east–west). The analyser position is above the objectives and can be slid in and out. When the analyser is in, this configuration is called cross-polarized light, with no light passing through the system (the view field is totally dark): this is abbreviated as XPL (cross-polarized light). If an anisotropic soil thin section is placed between the two polarizers, polarization colours appear due to different refractive indices (refringence) of objects in the soil. If the analyser is out, this configuration is abbreviated as PPL (plane-polarized light). The degree to which edges and surface imperfections of crystals is visible in PPL is called "relief". Moreover, linear traces can sometimes be observed through a mineral section in PPL or XPL: they are cleavage traces. Their angular relationships are used to identify some minerals.
3. Principle of mineral extinction: the intensity of the polarization colours varies during rotation of the stage from maximum brightness (1) to zero (extinction; 4). This succession is cyclic when rotating the stage 360°. Whenever the specimen is in extinction, the permitted vibration directions of light passing through are parallel with those of either the polarizer or analyser. However, if the thin section material is isotropic, it remains totally dark (extinction) when the stage is rotated through 360°.

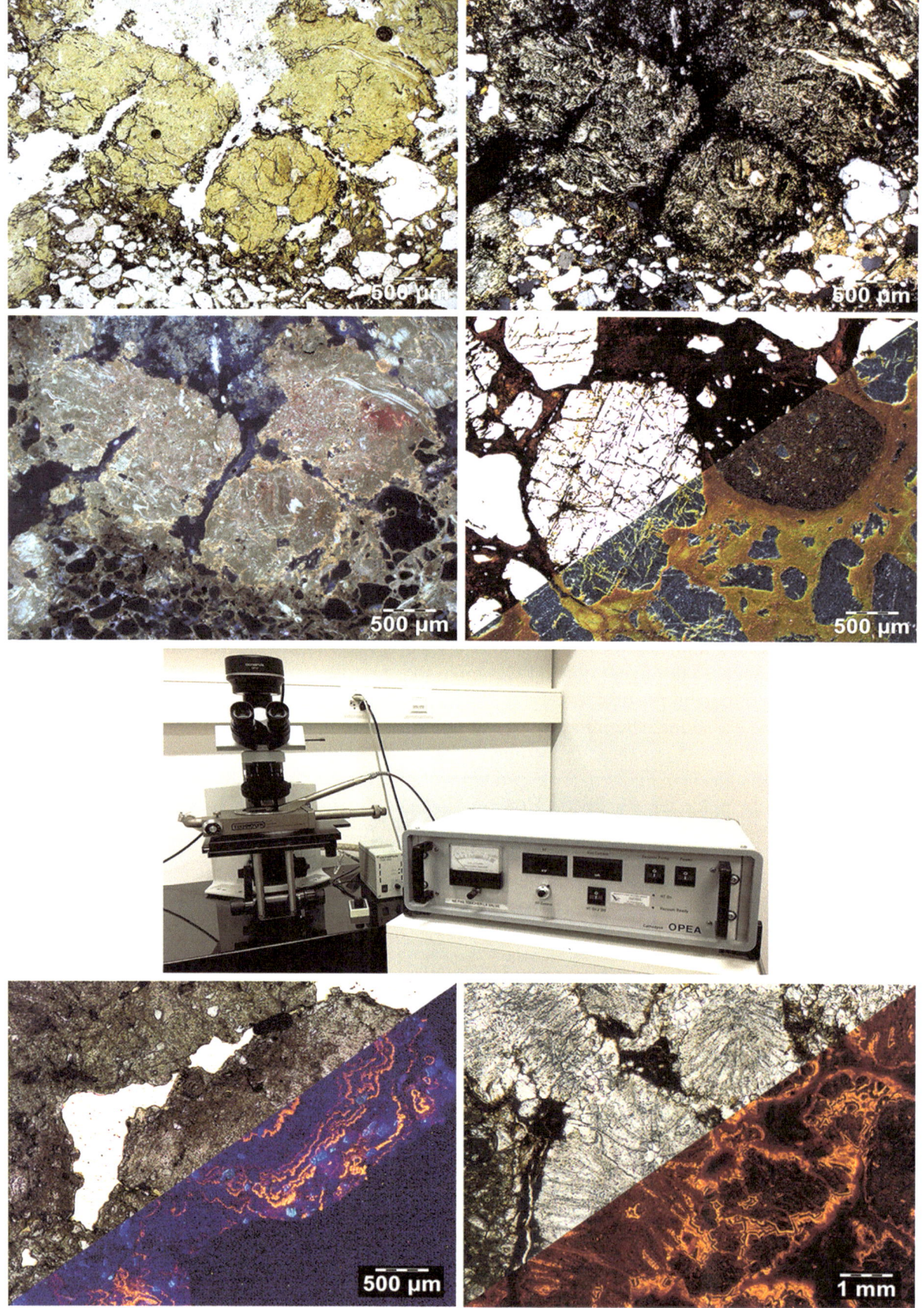

File 6: Other Techniques of Observation

> In soil micromorphology, not only are observations made in transmitted light but incident light is also used. The two most common techniques refer to fluorescence and cathodoluminescence. The source of light in fluorescence is usually an intense high-pressure mercury lamp ranging from 50 to 250 W. A high energy electron beam is the source of excitation in cathodoluminescence at a voltage between 10 to 20 kV. This equipment needs adaptation for the polarized light optical microscope.

Captions from upper left corner to lower right corner.

1. Yellow aggregates containing organic matter and phosphate from a dung deposit. Bottom part of the photograph: quartz sand in a dark and grey micromass. PPL.
2. Same view as 1. in XPL. Organic matter and phosphate remain extinct.
3. Same view as in items 1. and 2. under fluorescence blue-exciting radiation. The phosphate compounds appear in yellow, and the quartz grains are extinct (black). Some organic matter is excited and appears in bright pale blueish colours.
4. View in oblique incident light (OIL; lower right corner). OIL is a technique using direct reflected light projected on the thin section. It is widely used in archaeology (Goldberg and Macphail 2003; Nicosia and Stoops 2017). Dark greyish brown hematite-rich nodule surrounded by dark grey quartz in a micromass from a Ferralsol, Burkina Faso.
5. Optical microscope equipped with a cold-cathode cathodoluminescence (CL) device (University of Lausanne). A CL-stage, in which the sample is surrounded by an ionized gas in a moderate vacuum, is attached to the microscope. The electron beam used to bombard the thin section with high-energy electrons is generated by the discharge taking place between the cathode and the anode, at ground potential. The energy carried in the beam interacts with the crystal matter, resulting in the luminescence of minerals (Marshall 1988; Pagel et al. 2000; Boggs and Krinsley 2006).
6. Upper left corner: a carbonate nodule (PPL). Calcite appears in various brown tones with some small whitish quartz grains. Lower right corner: view in CL emphasizing multiple and alternating reddish to yellowish layers due to the presence of Mn^{2+} in the crystal lattice. The purple-blue luminescence is due to the intrinsic luminescence of Mn^{2+}-poor calcite (Richter et al. 2003).
7. Upper left corner: *Microcodium*, an unexplained feature currently observed in Tertiary paleosols developed in floodplains (PPL). Lower right corner: view in CL showing the structure of *Microcodium* in dark orange. The light orange rims are related to late diagenetic recrystallization, the luminescence being bright at a moderately high Mn^{2+} content, if Fe^{2+} abundance is below about 2000 ppm (Boggs and Krinsley 2006).

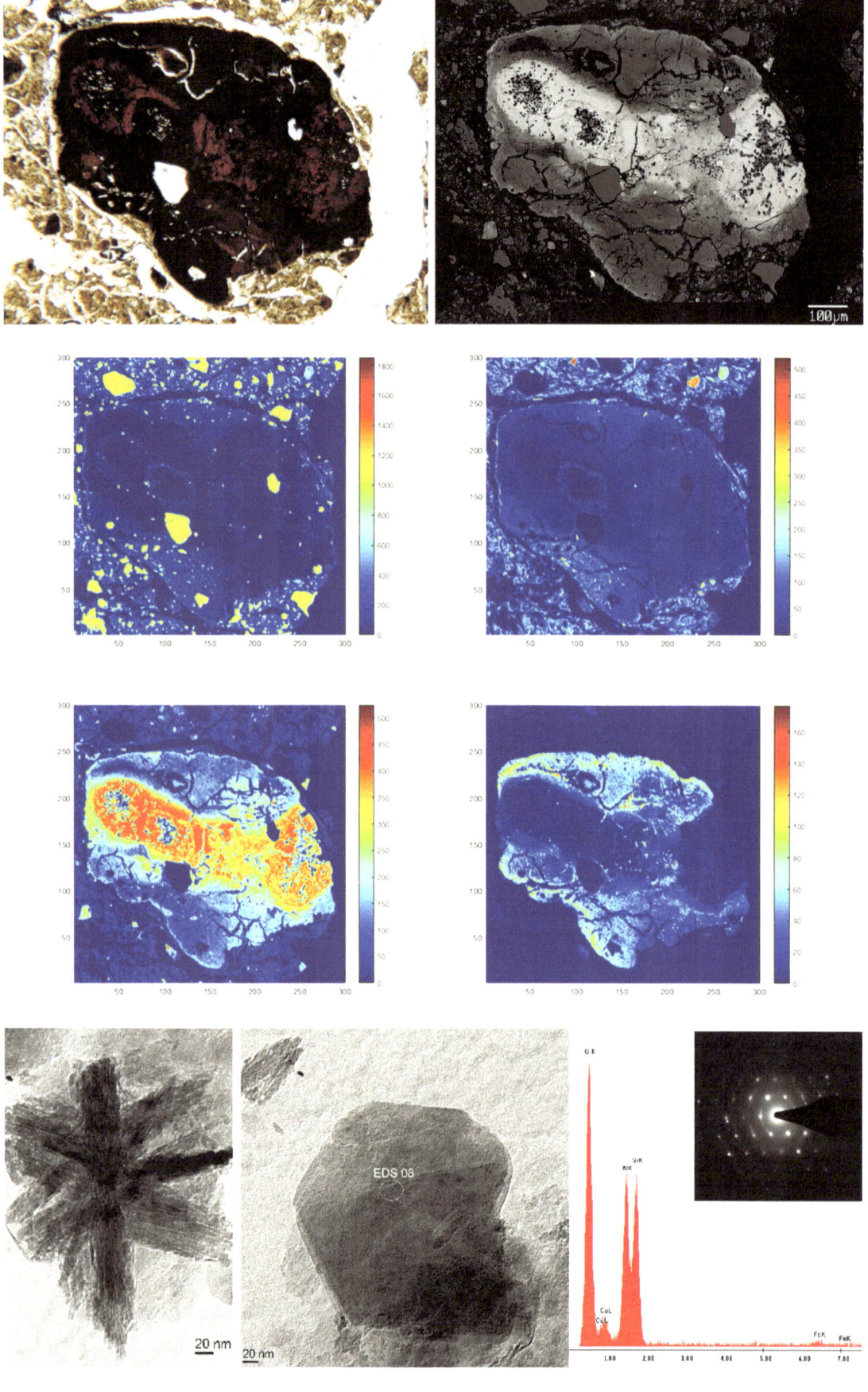

File 7: Electron and Energy Imaging

Soil micromorphologists can use the scanning electron microscope (SEM) to create images of features observed in thin sections. Not only can they access images at high resolutions, in order to see the minute structure of minerals and organic matter, but it is also possible to get information regarding the chemical composition of features. Electron probe micro-analyses (EMPA) are commonly performed to generate maps of chemical element distributions, whereas, transmission electron microscopes (TEM) are usually used to observe e.g. the structure of clay minerals or oxyhydroxides (see "File 78").

Captions from upper left corner to lower right corner.

1. Nodule composed of oxyhydroxides in a Cambisol developed on loess (Jura Mountains, Switzerland). The mineral fraction around the nodule includes some quartz grains and clay minerals (view in PPL).
2. Same nodule observed with an EMPA using the backscattered electron detector (BSE mode). The number of backscattered electrons reaching the detector is proportional to the mean atomic number of the sample: therefore, the light grey area correlates with heavier atoms, i.e. Fe and Mn, whereas the darker areas denote lighter elements (e.g. Al or Si).
3. Distribution map of Si: quartz grains appear in yellow or light green. They have high Si contents. The fine compounds around the nodule are enriched in Si compared to the nodule, which is highly depleted in Si.
4. Distribution map of Al: this map emphasizes the clay content of the fine fraction (light blue and green) around the nodule. The combination of Si and Al confirms the presence of aluminosilicates, in this case phyllosilicates.
5. Distribution map of Fe: the nodule contains a very high proportion of Fe in its centre, with a decreasing gradient towards its border.
6. Distribution map of Mn: the nodule periphery is enriched in Mn, whereas its centre is extremely depleted, as well as the fine fraction of the soil.
7. TEM view of a hematite crystal observed in the same Cambisol.
8. TEM view of an inherited crystal of kaolinite showing its hexagonal shape. At this magnification, it is possible to observe some superposition of individual sheet crystals. The "EDS 08" label refers to the spot analysed by EDS in 9.
9. In red, a spectrum of an EDS (Energy Dispersive X-ray Spectroscopy) analysis of the kaolinite crystal observed in 8. The spectrum shows the presence of Al, Si, and O, as expected. The presence of Cu is due to some contribution from the grids on which the sample is deposited. In the right upper corner, an image of the beam's electron diffraction pattern obtained for the kaolinite crystal. In order to get such images, the sample is tilted using a goniometric stage attached to the TEM.

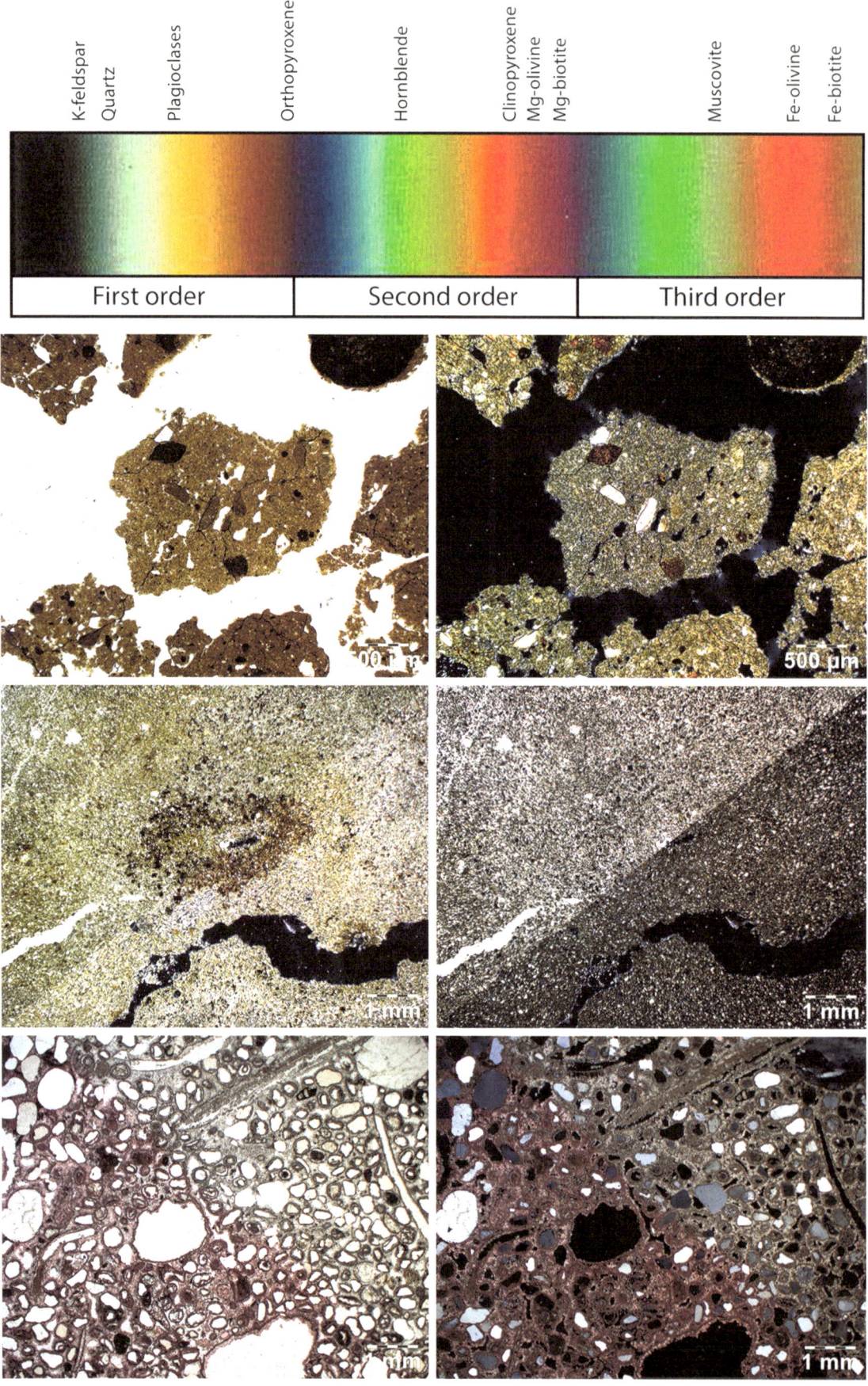

File 8: Colours of Minerals

When using a polarized microscope, light, which vibrates in a single plane, passes through the anisotropic minerals of the thin section (see "File 5"). Then, it splits into two beams perpendicular to each other, as it crosses the minerals, and propagates at different speeds according to the two refractive indices. The two vibrations emerge out of phase and pass through the analyser, which leads to the disappearance of certain wavelengths and to a resultant one, which defines the polarisation colour.

Captions from upper left corner to lower right corner.

1. In a 30 μm thick thin section, mineral colours are a function of the difference between the maximum and minimum indices of refringence, a property called birefringence. It depends on the orientation of the cross-polarized plane with respect to the crystalline system of the mineral. The birefringence varies with the rotation of the stage, having four intensity maxima and four extinction positions (see "File 5"). The colour chart, taken from a Michel-Lévy table of birefringence, covers different orders, from left to right, according to increasing retardation. Retardation refers to the distance that the slow beam lags behind the fast one by the time that the slow beam finally exits the mineral. Common inherited minerals observed in soils are plotted alongside the chart in relation to their maximal polarisation colours of their highest birefringence.

2. View in PPL of aggregates. Pores, as well as some minerals, appear in white.

3. Same sample as in item 2. in XPL. In the pore void, there is no retardation, so no light passes through the analyser, this area staying extinct. When minerals are clear in PPL and extinct in XPL, regardless of the orientation of the stage, they are called isotropic. Organic matter is often isotropic. Conversely, some minerals can remain white in XPL, a first order colour due to slow retardation, whereas other grains shift from very dark brown to light orange.

4. Oxyhydroxide accumulation around a pore. There are also oxyhydroxide and calcium carbonate impregnations in the groundmass, giving a yellowish to brownish colour to the sample.

5. Same sample as in item 4. after selective extraction of $CaCO_3$, Fe, and Mn. Such methods allow b-fabric, clay coatings, and microcrystalline features to be enhanced. It can also reveal the nature of the parent-material groundmass.

6. Recent soil developed on a siliclastic and carbonate sediment (PPL). The lower left triangle has been soaked in a solution containing alizarine red and potassium ferricyanide (Stoops 2003, 2021), staining high Mg-calcite red.

7. Same sample as in item 6. in XPL. Quartz grains appear in white, grey, and extinct, calcite in red (lower left triangle) or greyish brown (not stained; upper right triangle).

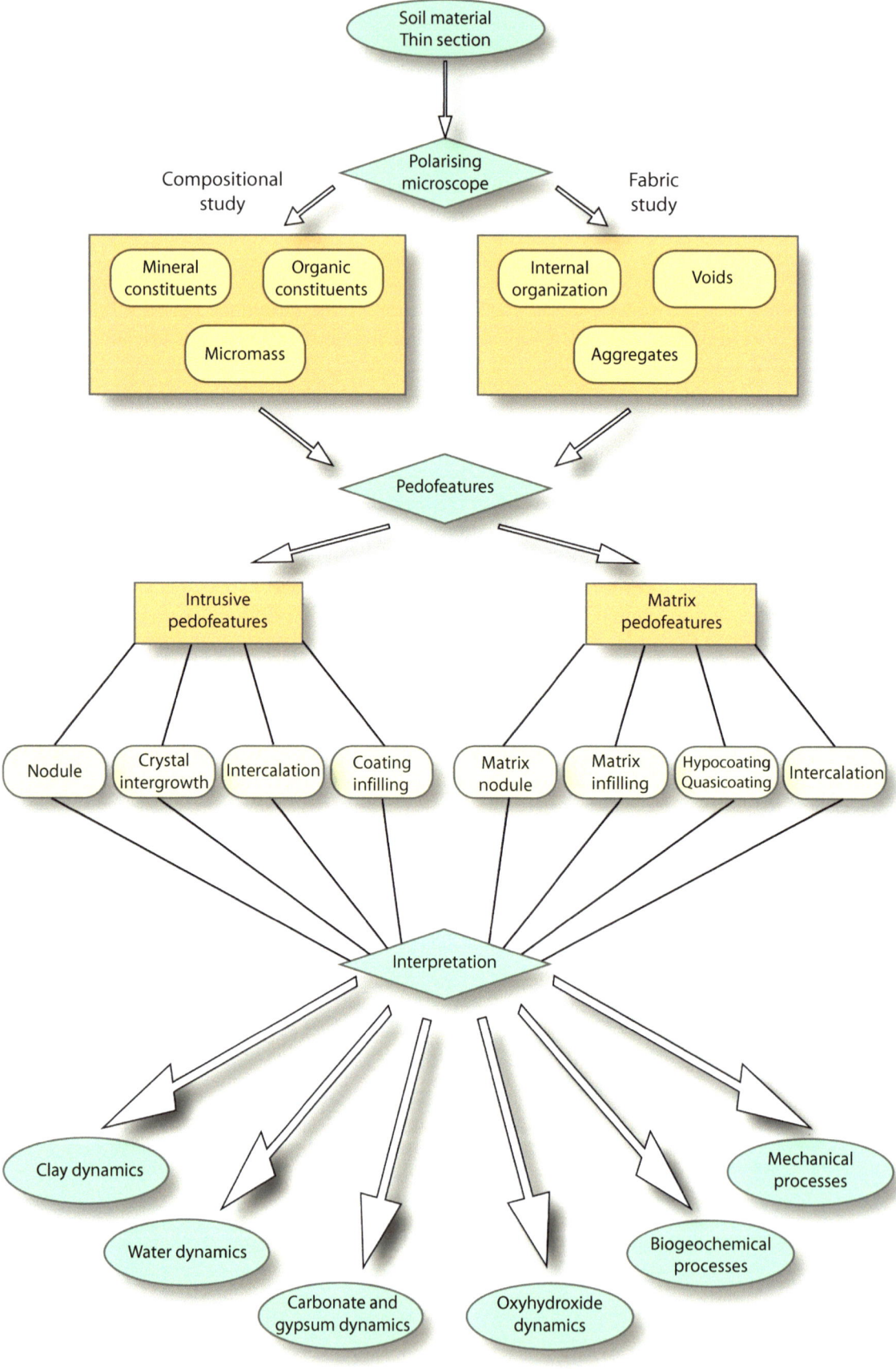

File 9: The Micromorphological Approach

> The micromorphological approach is based on multiscalar observations of both composition and fabric. The chart is derived from the approach proposed by Stoops (2003). The oval boxes refer to input and output of the flow, the diamonds to decisions, rectangles to an identification process, and rounded rectangles to objects.

Captions from top to bottom.

1. The soil material is sampled in the field (see "File 3") and the thin section is prepared (see "File 4").
2. The study starts with the investigation of the groundmass using a polarizing microscope (see "File 5"). The groundmass is the base material of the soil in thin section. Two different branches of investigation are carried out to study this groundmass: (1) the compositional study (see Chap. 3) dealing with "data such as chemical and/or mineralogical composition or associated characteristics such as colour, refractive index, or interference colours" (Stoops 2003); the studied objects include mineral constituents, organic constituents, and micromass; and (2) the fabric (see Chap. 2), which focuses on the microstructure, aggregates, and voids.
3. After the descriptions of the various components and fabric of soil thin sections, the next step consists of the identification and description of pedofeatures (see Chap. 4). Pedofeatures are "discrete fabric units present in soil materials that are recognizable from adjacent material by a difference in concentration in one or more components or by a difference in internal fabric" (Bullock et al. 1985).
4. There are two main groups of pedofeatures: the matrix pedofeatures and the intrusive pedofeatures (Stoops 2003, 2021). The former refers to changes within the groundmass. The latter deals with changes outside the groundmass.
5. Each of the two groups of pedofeatures is subdivided into different subgroups, according to their morphology (Stoops 2003, 2021).
6. All the information gathered is used to interpret the thin section in terms of soil genesis, soil classification, soil use, palaeopedology, and geoarchaeology. However, the pedogenetic processes are generally not defined on a morphological basis, but mainly on chemical, physical, and biological properties and by using comparisons between profile horizons. Nonetheless, soil micromorphology is useful for the identification of the main dynamics affecting the soils.
7. Examples of six different dynamics that can be identified using soil micromorphology (see Chap. 5): the dynamics of clays, the impact of water, the precipitation of carbonates, sulphates, and chlorides, the formation of oxyhydroxides, the development of biogeochemical processes, e.g. organomineral interactions, and mechanical processes, e.g. gravity, compaction, or ploughing.

Chapter 2

The Organization of Soil Fragments

File 10: Concept of Fabric

Kubiëna (1938) was the first to introduce the concept of fabric in soil micromorphology, so this term has been used in soil micromorphology for a long time. The term "fabric" was initially applied to rocks by geologists and petrologists. This type of fabric is defined as the "factor of the texture of a crystalline rock which depends on the relative sizes, the shapes, and the arrangement of the component crystals" (Matthews and Boyer 1976). This definition has been adapted for soil micromorphology and its latest definition has been given by Bullock et al. (1985) as: "soil fabric deals with the total organization of a soil, expressed by the spatial arrangement of the soil constituents (solid, liquid, and gaseous), their shape, size, and frequency, considered from a configurational, functional and genetic view-point". In conclusion, the soil micromorphologist should consider the fabric as an arrangement and/or organization of soil constituents.

Fabrics can be very complex and this concept can be encountered in many different circumstances. For instance, the concept of fabric is mainly related to soil microstructure (see "File 9", "File 20", and "File 21"), but also associated to the c/f related distribution (see "File 13" and "File 14"), b-fabric (see "File 45" and "File 46"), as well as pedofeatures (see "File 9" and Chap. 4). Generally speaking, the fabric is related to the type of light used, as well as the scale of observation, i.e. the magnification of the microscope lens (see "File 11").

Examples of various fabrics related to the main c/f distributions (see "File 13"). All microphotographs are in PPL.

© The Author(s) 2021
E. P. Verrecchia, L. Trombino, *A Visual Atlas for Soil Micromorphologists*,
https://doi.org/10.1007/978-3-030-67806-7_2

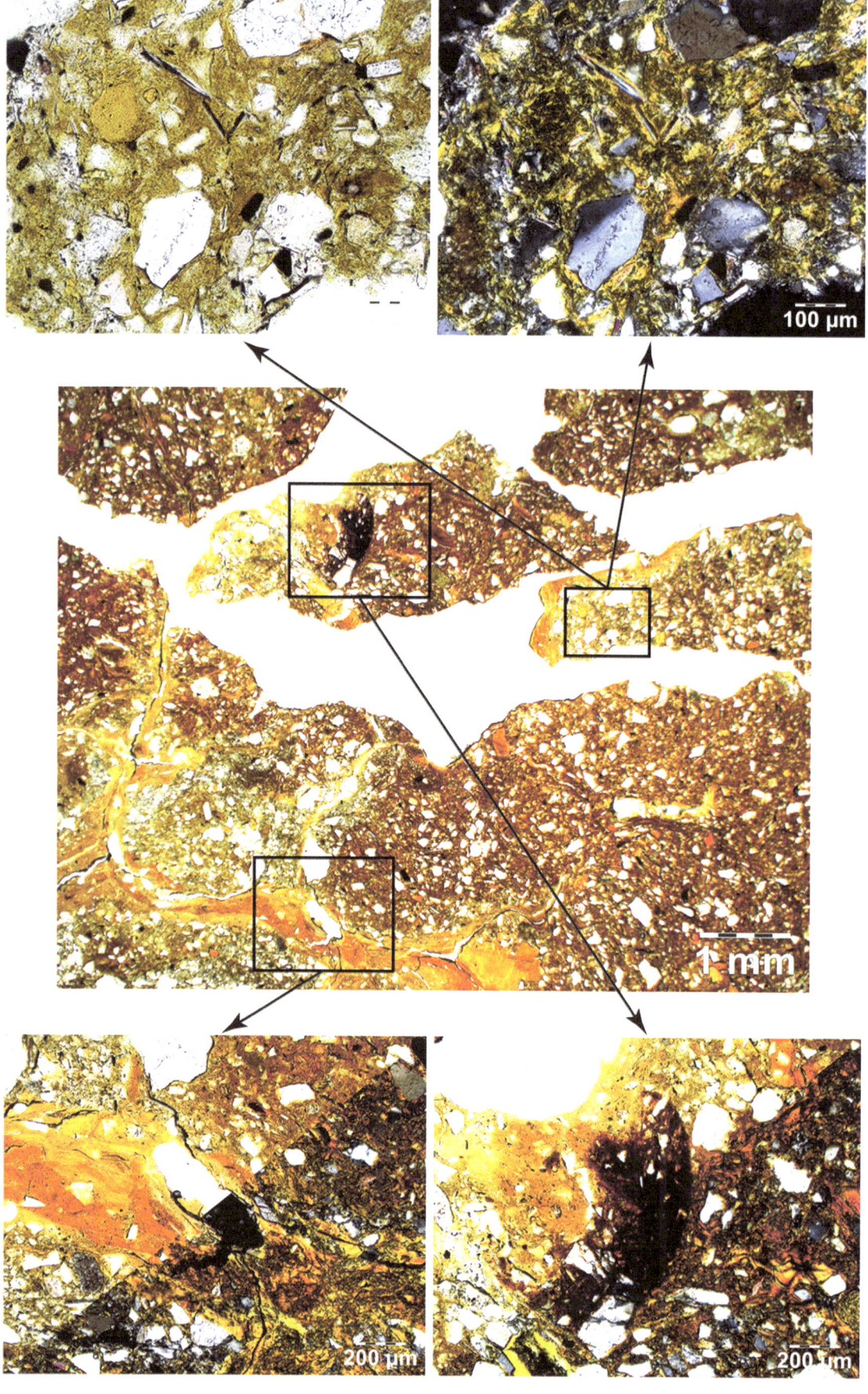

File 11: Multiscalar Approach to Fabric

Fabric is a multiscalar concept that is used to describe homogeneous and heterogeneous units. The example given in this section shows fabrics observed at various magnifications with different soil components and features. Fabric units are units delimited by natural boundaries, visually homogeneous at the scale of observation and distinct from other fabric units (Bullock et al. 1985; Stoops 2003). However, increasing the magnification leads to an increase in either homogeneity or heterogeneity, depending on the fabric and feature involved.

1., central view; 2.–3., upper views; 4.–5., lower views.

1. Central picture: general view at low magnification of a soil thin section (in PPL) showing a mottled soil groundmass with large elongated voids (in white). The arrows point to magnified details of the fabric displayed in the microphotographs.

2.–3. Close-up view showing the presence of quartz grains (each of them being a fabric unit) in a clayey and fine silty micromass (left: PPL; right: XPL). This arrangement constitutes the base material of the soil.

4. Left: a specific pedofeature called "clay coating" (see "File 56" and "File 57"), formed by clay layers associated with iron oxyhydroxides. These coatings display a specific fabric, also called "partial fabric" by Bullock et al. (1985) and Stoops (2003) at the given magnification.

5. Right: a specific pedofeature called a "matrix nodule" formed by the concentration of oxyhydroxides in the groundmass (see "File 48"). This nodule is characterized by a specific fabric at the given magnification.

File 12: Basic Distribution Patterns

A pattern is the spatial arrangement of fabric units (Stoops 2003). Two types of patterns are usually defined: the distribution patterns and the orientation patterns. This section illustrates the basic distribution patterns commonly observed in thin sections, which are the distribution of fabric units of the same type with regard to each other (Stoops 2003).

Captions from upper left corner to lower right corner.

1.–3.   Three examples of banded patterns formed by alternating layers of different grain size. (1) Alternating clayey, silty, and coarse layers (PPL); (2) Alternating coarse and silty layers; (3) Alternating very fine and fine clayey layers.

4.   Linear pattern illustrated by a fabric unit organized along a line.

5.   Random pattern characterized by a random distribution of the fabric units.

6.   Fan-like pattern constituted by multiple clay infillings, organized in a fan-like morphology. In this case, the fabric units create a fan-like pattern at a specific scale.

7.   Clustered pattern defined by a grouping of fabric units emphasized by both their colour and the void distribution in this example.

8.   Interlaced pattern showing interlaced fabric units constituted by clay coatings and infillings in this example.

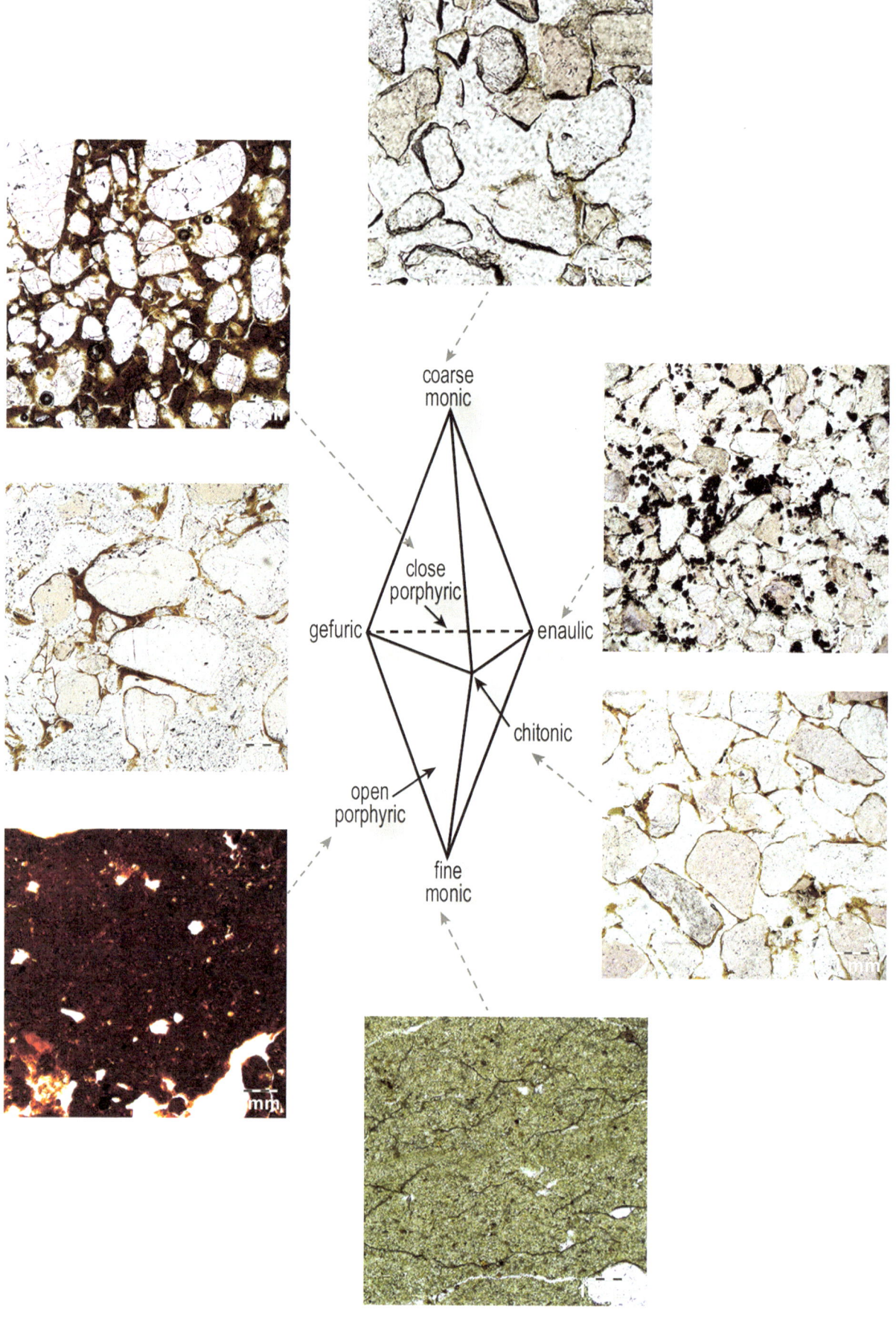

coarse
monic

close
porphyric

gefuric ──────────── enaulic

chitonic

open
porphyric

fine
monic

File 13: c/f Related Distributions I

> The c/f related distribution refers to the distribution of coarse fabric units compared to fine fabric units and, if applicable, their associated pores. It has to be emphasized that this concept is purely descriptive and does not consider the interpretation of such fabric units. Stoops and Jongerius (1975) proposed a bipyramid of tetrahedra to summarize the basic c/f related distribution. This geometrical shape is modified and used in this section to illustrate the main c/f patterns.

Captions given clockwise from the upper top picture. All microphotographs in PPL.

1. Coarse monic: there is only one size of fabric unit, in this case, which is coarser than the given c/f limit chosen by the observer. Quartz grains are loosely arranged with a quasi-absence of fine material.
2. Enaulic: fine fabric units form small aggregates (i.e. micro-aggregates) inside the space between the coarse fabric units. Organic micro-aggregates spread inside the space between quartz grains.
3. Chitonic: coarse fabric units are coated by fine fabric units. Quartz grains are surrounded by brownish clay and oxyhydroxide thin layers.
4. Fine monic: there is only one size of fabric unit, in this case, which is finer than the given c/f limit chosen by the observer. A very fine silty and clayey mixture with a quasi-absence of coarse material.
5. Open porphyric: the coarse fabric units are scattered in a dense micromass of fine fabric units, and the distance between the coarse fabric units is more than twice their average size. Rare quartz grains are floating in a reddish clayey micromass.
6. Gefuric: coarse fabric units are connected by bridges of fine fabric units. Quartz grains are bound by brownish clay and oxyhydroxide material.
7. Close porphyric: the coarse fabric units are dispersed in a dense micromass of fine fabric units, and have many points of contact. Quartz grains are embedded in a reddish clayey micromass.

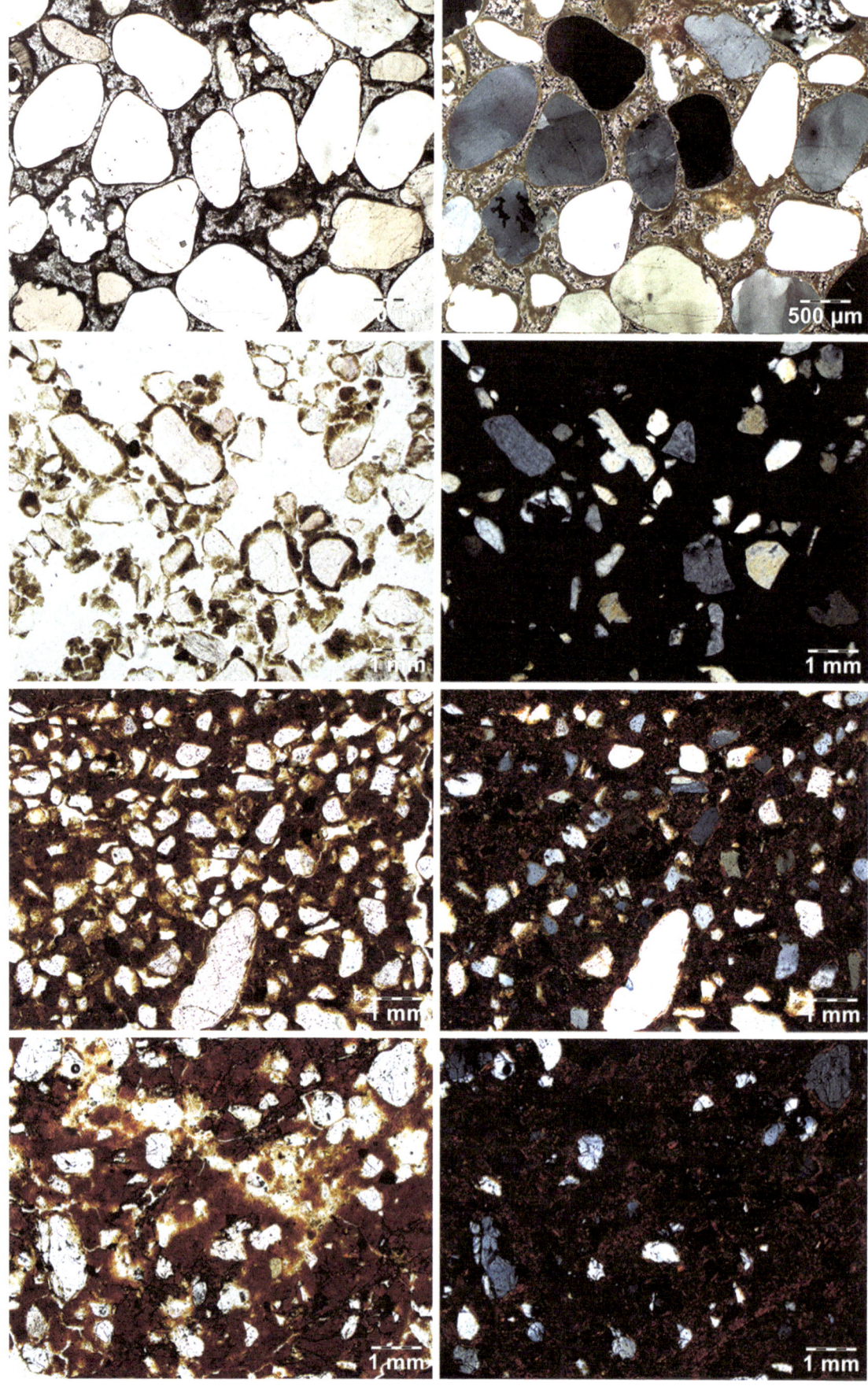

File 14: c/f Related Distributions II

> The c/f related distribution refers to the distribution of coarse fabric units compared to fine fabric units and, if applicable, their associated pores. It has to be emphasized that this concept is purely descriptive and does not consider the interpretation of these fabric units. This section shows different variations in chitonic, gefuric, and porphyric c/f related distributions.

Captions from upper left corner to lower right corner.

1.–2. Chitonic: coarse fabric units are coated by fine fabric units. In the picture, quartz grains are surrounded by thin layers of micritic calcite. The space between the grains is secondarily filled by a cement made of microsparitic calcite (left: PPL; right: XPL).

3.–4. Chito-gefuric: the coarse fabric units are both coated and connected by bridges of fine fabric units. Quartz grains are surrounded and bounded by brownish clay and fine material rich in oxyhydroxides (left: PPL; right: XPL).

5.–6. Single-spaced porphyric: scattered coarse fabric units float in a dense micromass of fine fabric units. However, the distance between the coarse fabric units is less than their average size. Quartz grains are dispersed in a reddish clayey micromass (left: PPL; right: XPL).

7.–8. Double-spaced porphyric: the coarse fabric units are floating in a dense micromass of fine fabric units, and the distance between them is more than one and less than twice their average size. Quartz grains are embedded in a reddish clayey micromass (left: PPL; right: XPL).

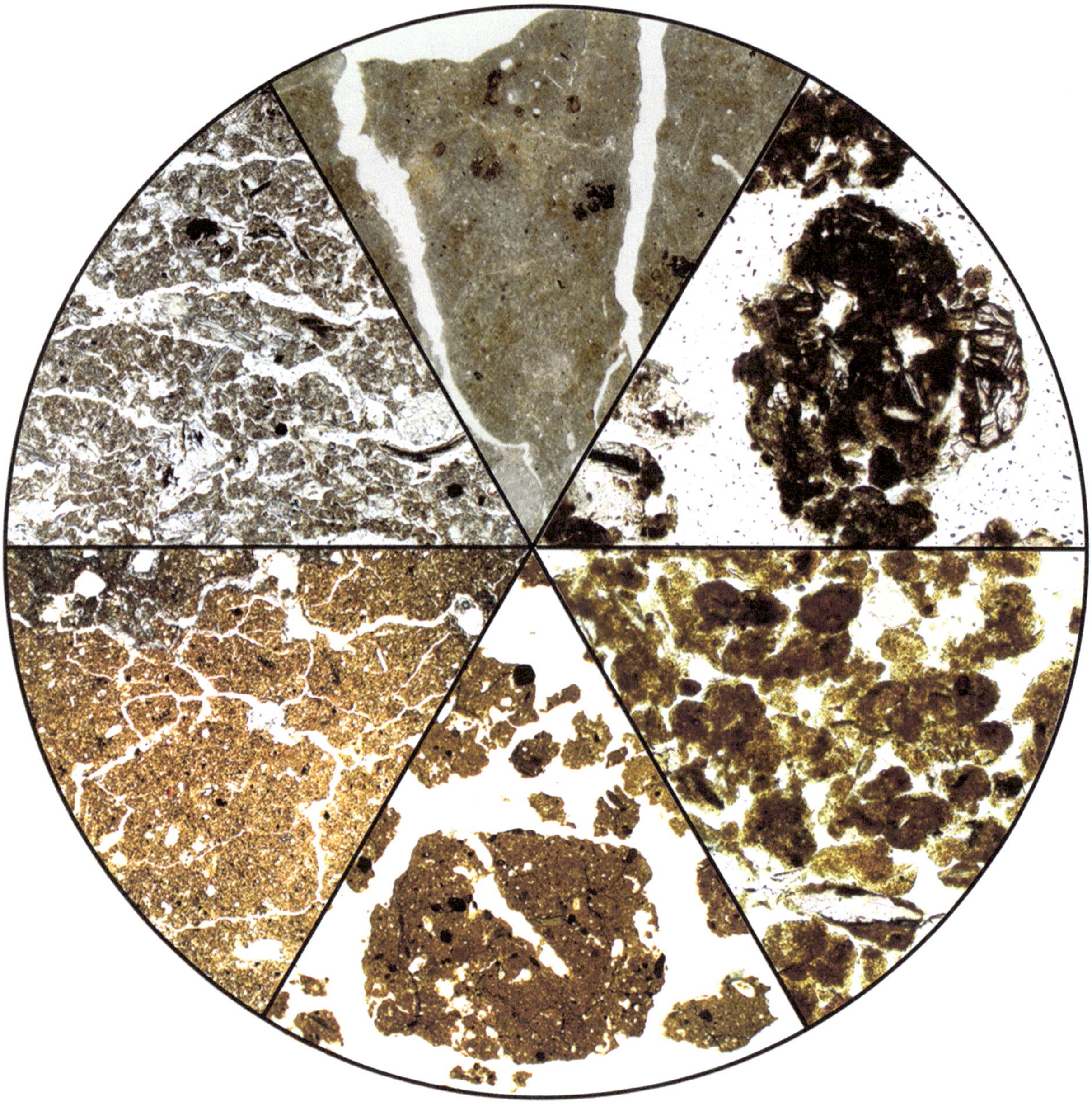

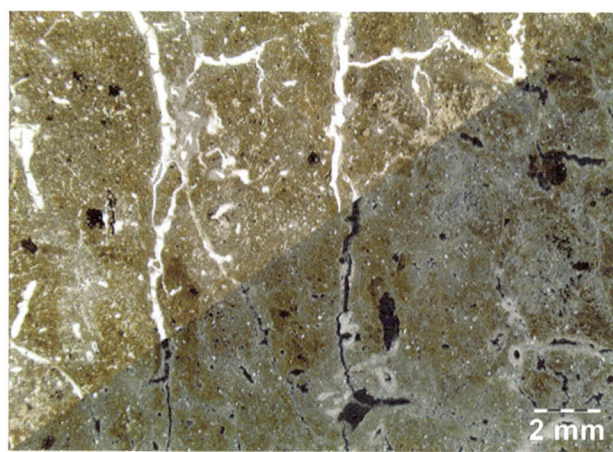

File 15: Aggregates and Aggregation

Aggregation and aggregates (also called peds) are directly related to the soil structure. Their role is fundamental in defining soil properties, as aggregates are typically a product of pedogenesis. Aggregates are bodies separate from the soil groundmass, clearly delimited from each other and/or the surrounding soil material. In soil microscopy, they are first defined by their morphology.

Captions from top circle clockwise to lower right corner.

1. Captions are given clockwise starting from the uppermost slice. Photographs are in PPL. *Prisms*: they are vertically elongate aggregates, usually bounded by planar voids; prisms are only clearly visible if the thin section is big enough to contain them (see item 3., this plate). *Crumbs*: variously rounded peds appearing porous at the microscopic scale. *Granules*: variously rounded peds appearing non-porous at the microscopic scale. *Subangular blocky peds*: peds of more or less equant size with a subangular shape; they often correspond to a macroscopic subangular blocky structure. *Angular blocky peds*: peds of more or less equant size with angular edges; they mostly correspond to a macroscopic polyhedral structure. *Plates*: plates form generally elongate and sub-horizontal aggregates; in the picture, plates have a lenticular shape, but it is also possible to observe simple straight plates.
2. Angular blocky aggregates: at this low magnification, it is possible to see the general shape of the angular large peds.
3. Prisms at low magnification: the network of vertical and horizontal planar voids clearly separates the prismatic aggregates.

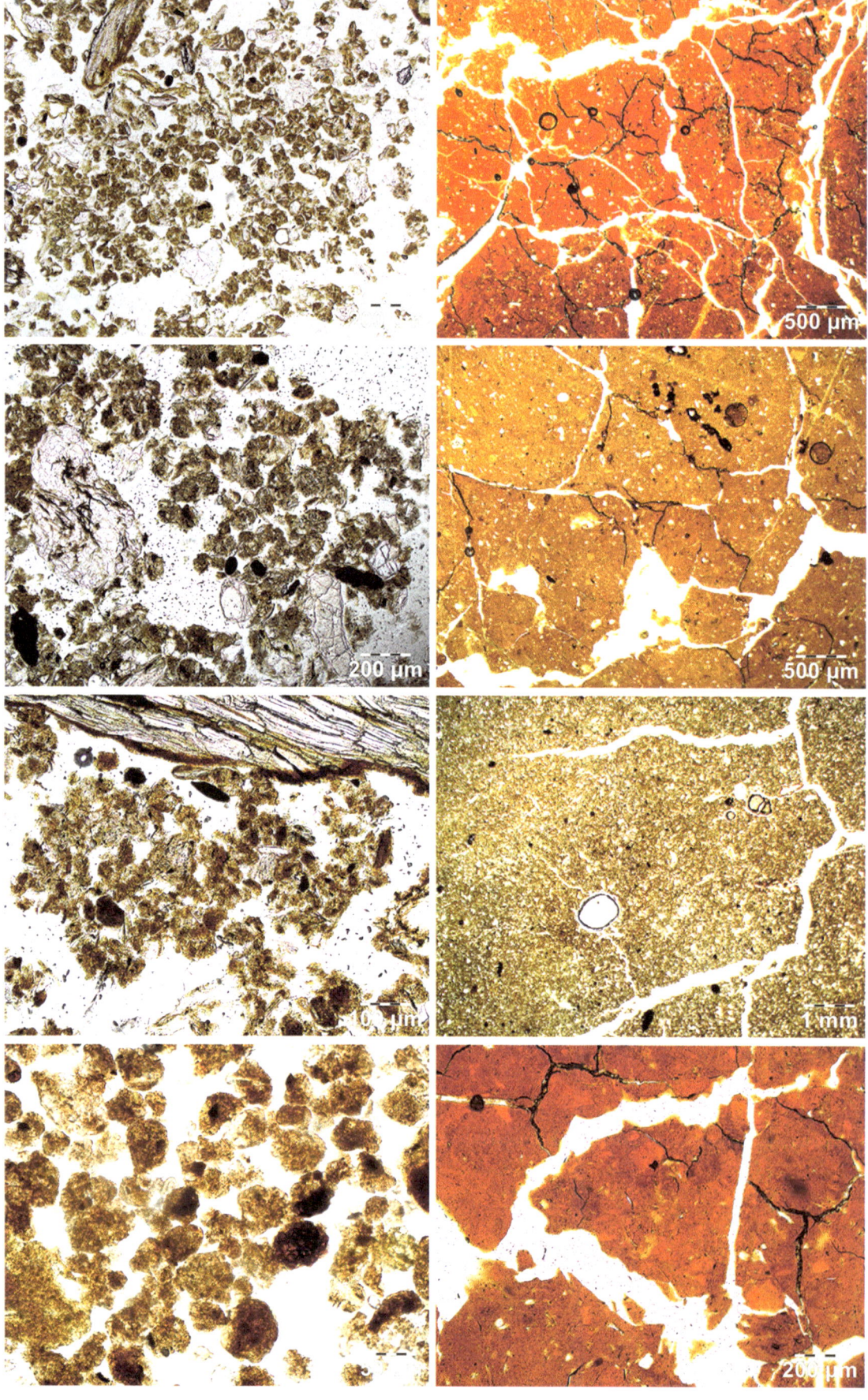

File 16: Degree of Separation and Accommodation of Aggregates

Aggregation and aggregates (also called peds) are directly related to the soil structure. Their role is fundamental in defining soil properties, as aggregates are typically a product of pedogenesis. Aggregates are bodies separate from the soil groundmass, clearly delimited from each other and/or the surrounding soil material. In addition to their morphology, they are also defined by their degree of separation and accommodation. The degree of separation refers to preferential zones of weakness illustrated by voids in soil microscopy. The accommodation is the degree to which adjacent ped faces coincide in a complementary way.

Captions from upper left corner to lower right corner. All microphotographs in PPL.

1. Granular aggregates showing a high degree of separation.
2. Angular blocky peds with a high degree of separation.
3. Crumb aggregates with a moderate degree of separation.
4. Angular blocky peds with a moderate degree of separation.
5. Crumb aggregates with a weak degree of separation.
6. Weakly separated angular blocky peds.
7. Rounded aggregates cannot be accommodated.
8. Example of accommodated blocky aggregates.

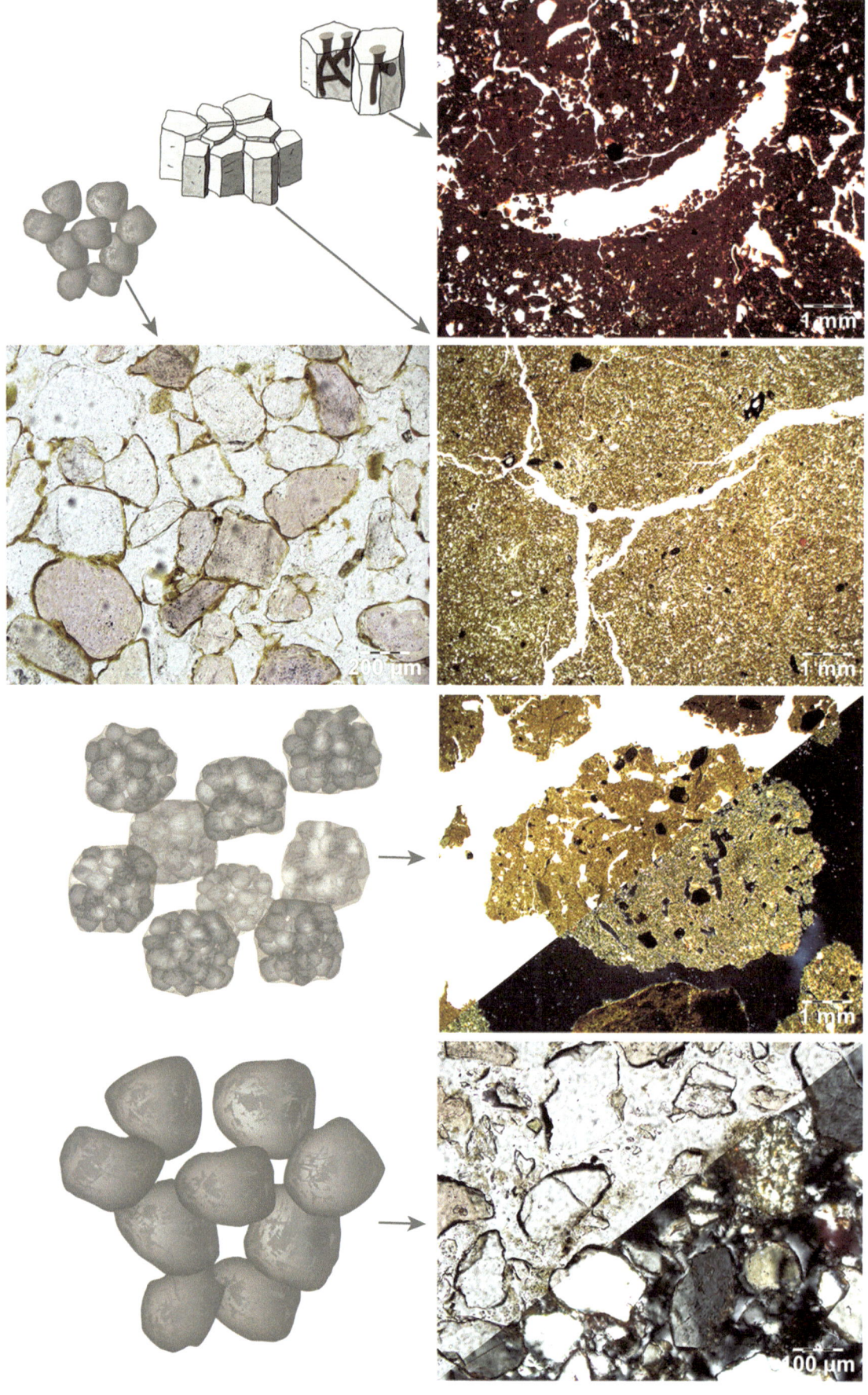

File 17: The Nature of Voids

> Voids are spaces unoccupied by soil material. Soil micromorphologists distinguish between various types of voids, according to their shape and arrangement. Moreover, in soil microscopy, voids between clay particles are not taken into account as they are below the resolution of the optical microscope.

Captions from upper left corner to lower right corner.

1. Soil scientists usually describe the nature of pores using a typology based on their origin, i.e. resulting from (a) packing of primary soil particles (bottom left), (b) cracks separating structural peds (centre), and (c) biological activity (upper right corner). Sketch modified from Weil and Brady (2017).
2. Example of a biological pore: this type of pore is called a "channel" (PPL).
3. Simple packing voids between quartz grains coated by fine material (PPL).
4. Planes, i.e. planar voids between angular peds (PPL).
5. In some soil aggregates, two types of pores have to be recognized: the intrapedal and the interpedal pores. Sketch modified from Weil and Brady (2017).
6. Subangular blocky ped with intrapedal pores surrounded by compound packing voids.
7. When the soil material is mainly composed by individual mineral particles, voids are only composed by the space between the particles themselves. Sketch modified from Weil and Brady (2017).
8. Quartz grains surrounded by simple packing voids.

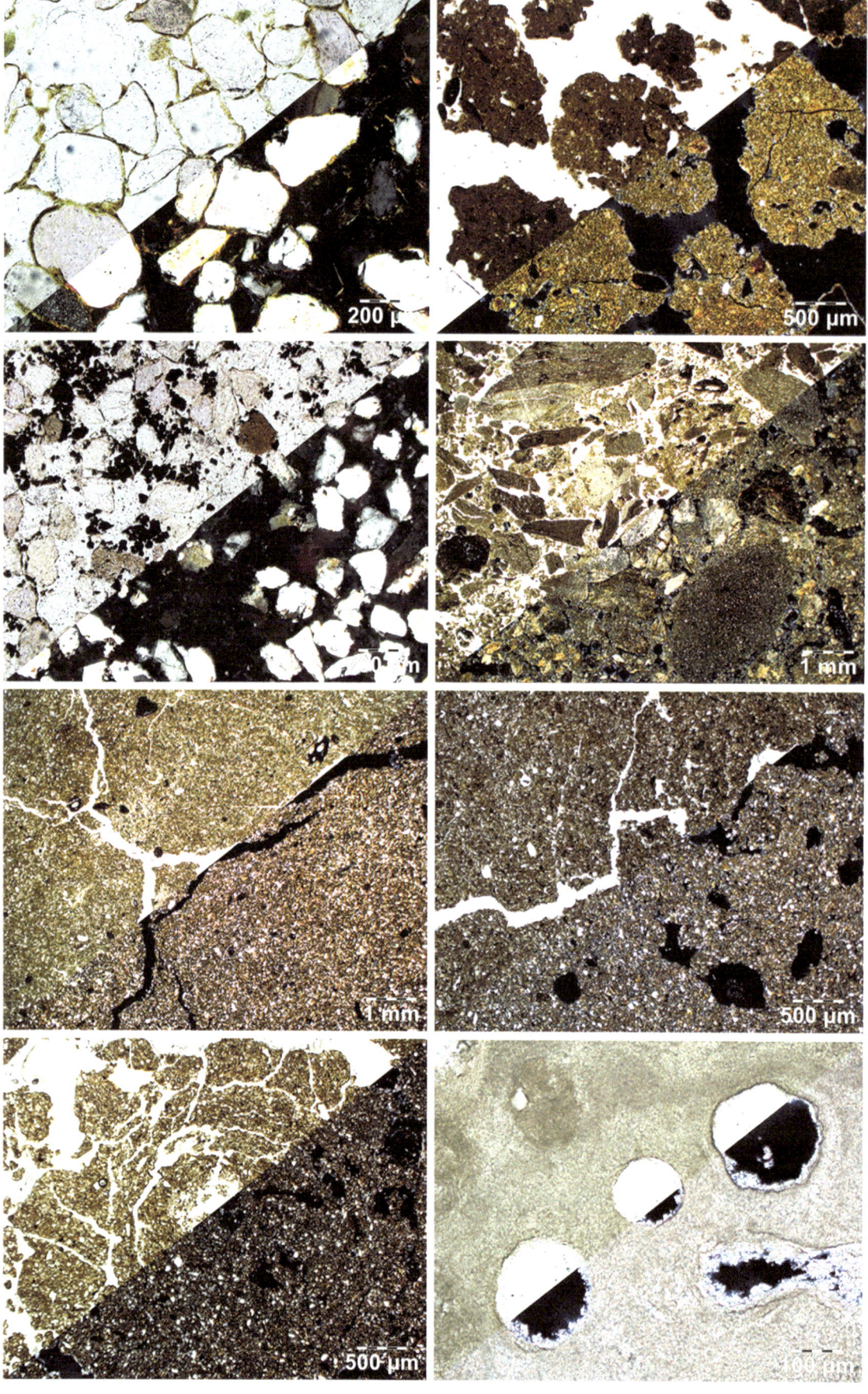

File 18: Morphology of Voids I

> Voids are spaces unoccupied by soil material. Soil micromorphologists distinguish between various types of voids according to their shape and arrangement. Once their nature is identified, voids can be classified according to their morphology.

Captions from upper left corner to lower right corner.

1. Simple packing voids separating quartz grains coated by fine material.
2. Compound packing voids between non-accommodated crumb to subangular peds.
3. Complex packing voids separating quartz grains and organic micro-aggregates.
4. Complex packing voids between coarse rock fragments and soil aggregates of different sizes.
5. Straight planes are flat voids, accommodated with sharp pointy ends.
6. Zigzag planes are cracks abruptly changing directions throughout their length.
7. Curved planes are curved to circular voids, sometimes occurring as an onion skin with multiple layers.
8. Vesicles are round voids often observed in groups. They commonly indicate the presence of air bubbles in the soil material.

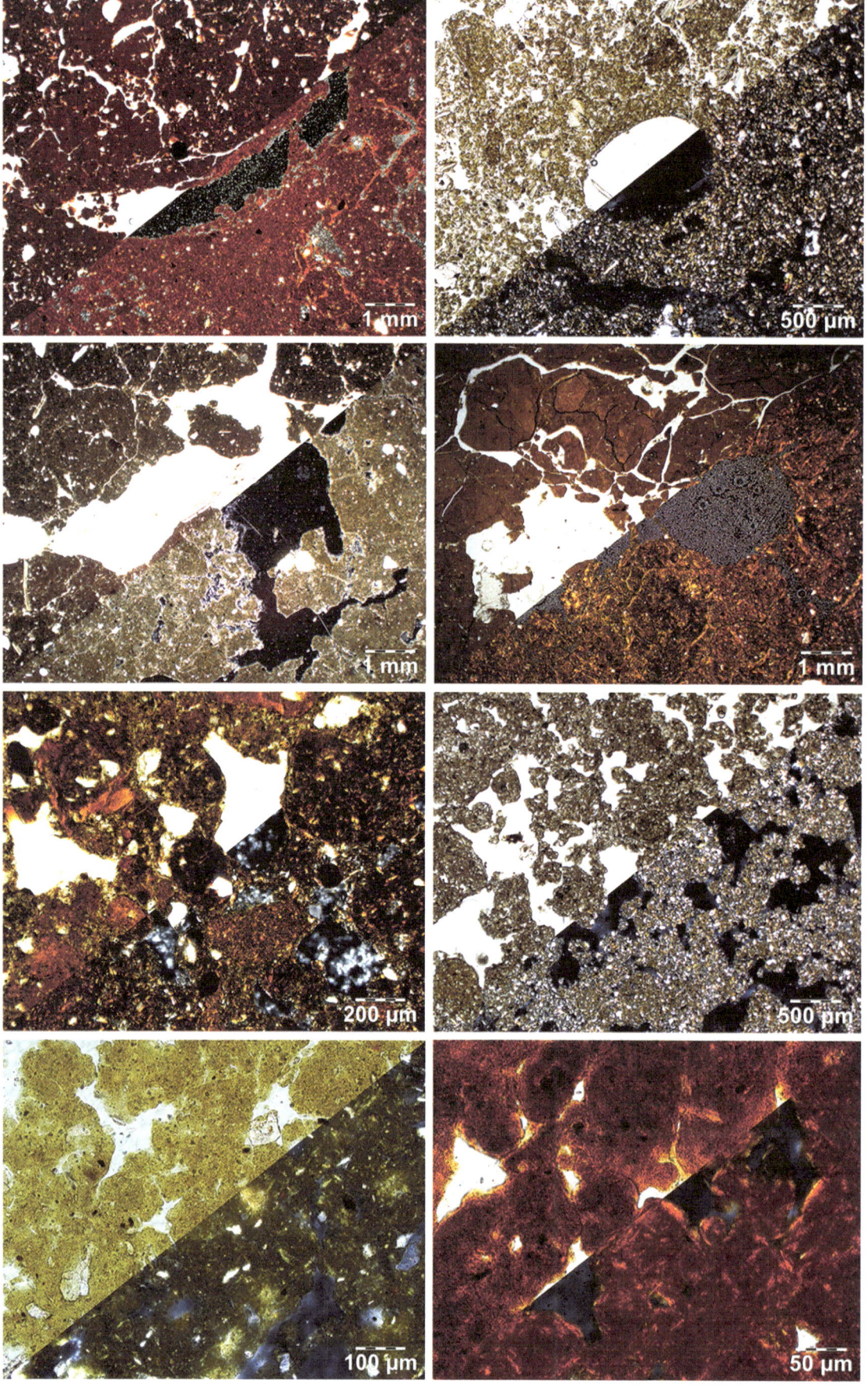

File 19: The Morphology of Voids II

> Voids are spaces unoccupied by soil material. Soil micromorphologists distinguish between various types of voids, according to their shape and arrangement. Once their nature is identified, voids can be categorized according to their morphology.

Captions from upper left corner to lower right corner.

1. Longitudinal section of a channel: such voids are tubular and smooth and cylindrical or arch-shaped in section. They are fairly uniform all along their length.
2. Transversal section of a channel: similar to the void described above, the section is usually round and smooth and may show some evidence of compacted soil material around the edges.
3.–4. Chambers are specific voids characterized by interconnecting channels and planes. Generally speaking, they are often larger than other voids in the same thin section.
5.–6. Vughs are irregular and non-elongated voids.
7.–8. Star-shaped vughs refer to voids having a triangular to polygonal shape due to the welding of rounded aggregates.

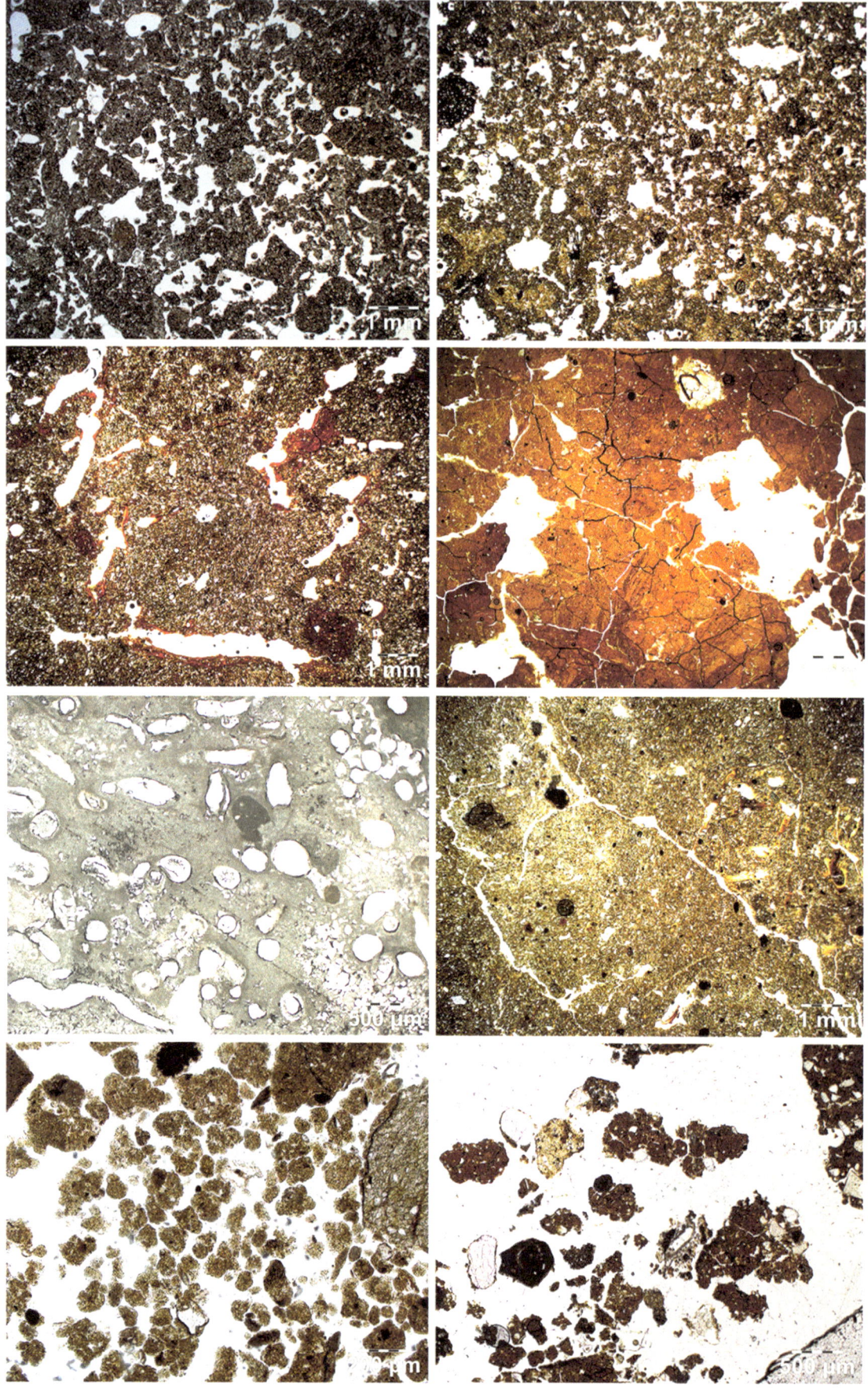

File 20: Microstructure I

> Microstructure is a term used to describe the relationship between the solid and the non-solid phases of the soil. It is defined using the morphologies of aggregates and voids, degree of separation of aggregates, as well as the relationships between voids, aggregates, and mineral grains. It is obvious that soils display different types of microstructure according to the magnification used during observation. Therefore, a choice must be made for the description, as this concept is supported by a comprehensive approach to the thin section.

Captions from upper left corner to lower right corner. All microphotographs in PPL.

1. Vughy microstructure: dominant vughs disrupt the continuity of soil fine material.
2. Spongy microstructure: whatever the type of voids, they are numerous and often interconnected, interrupting the continuity of soil material.
3. Channel microstructure: an abundance of channels is present inside the soil groundmass.
4. Chamber microstructure: an abundance of chambers is present inside the soil groundmass.
5. Vesicular microstructure: an abundance of vesicles is present inside the soil groundmass. Some of the pores are also channels in the given example.
6. Fissure/crack (Bullock et al. 1985) microstructure: blocky aggregates are not fully separated. Generally, the groundmass is dense except for a few planes and possibly some channels, as shown in the picture.
7. Granules: small, rounded, non-accommodating aggregates, with no noticeable internal porosity. They are separated by compound packing voids and, in this case, the microstructure is termed as granular.
8. Crumbs: aggregates made of small, even tiny, clusters of groundmass, unevenly welded together, often more or less rounded, and incorporating some small pores. Put together, crumbs form a crumb microstructure.

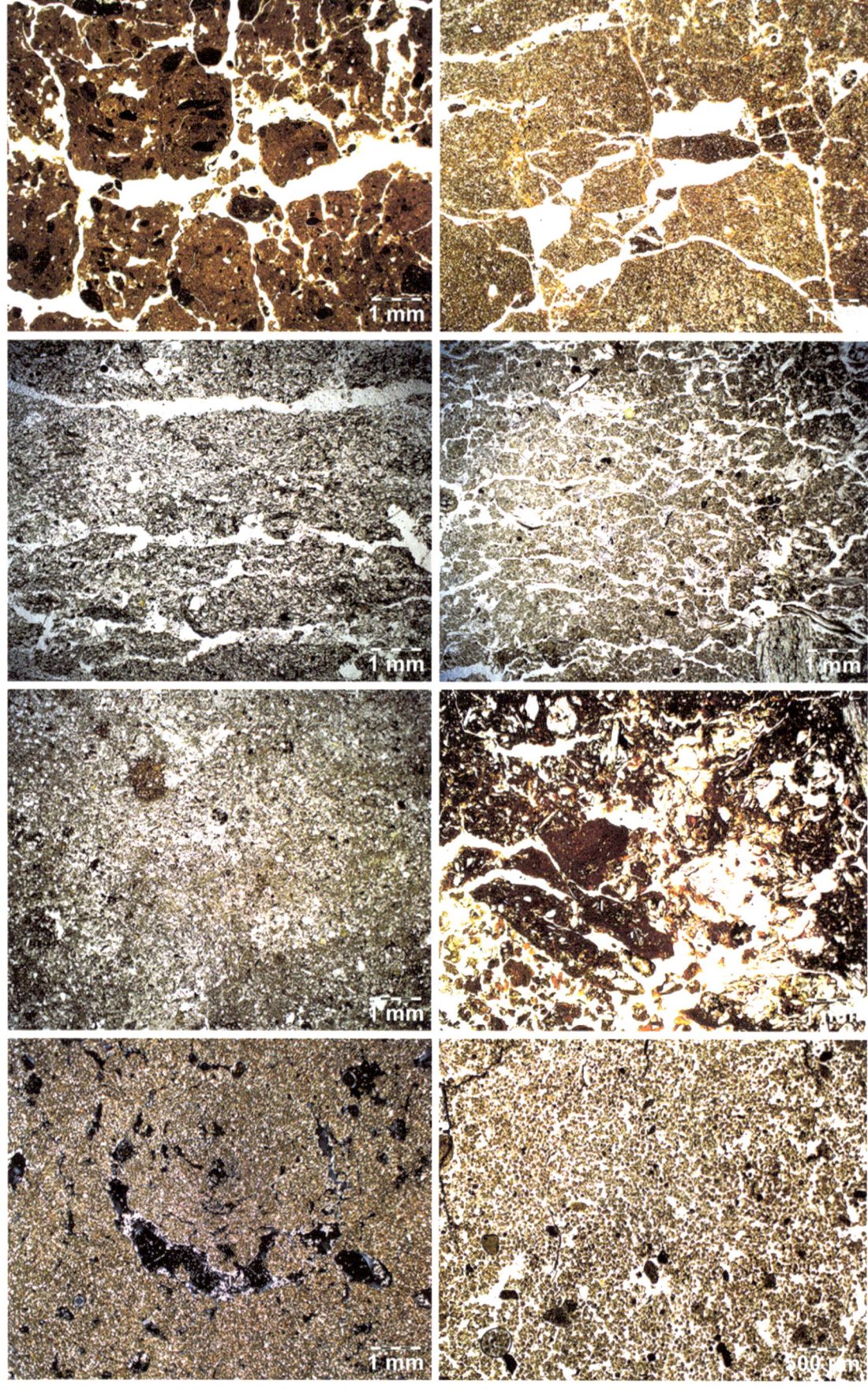

File 21: Microstructure II

> Microstructure is a term used to describe the relationship between the solid and the non-solid phases of the soil. It is defined using the morphologies of aggregates and voids, degree of separation of aggregates, as well as the relationships between voids, aggregates, and mineral grains. It is obvious that soils display different types of microstructure according to the magnification used during observation. Therefore, a choice must be made for the description, as this concept is supported by a comprehensive approach to the thin section.

Captions from upper left corner to lower right corner. All microphotographs in PPL if not otherwise specified.

1. Subangular blocky microstructure: short planar voids separate subangular blocky peds on all or most of their sides. Some other voids, such as small channels or vughs, can also be present inside or between aggregates. Finally, the faces of aggregates accommodate each other well.
2. Angular blocky microstructure: aggregates with sharp angular edges. Aggregates are separated by a network of planar voids and accommodate each other well. Other types of voids are rare.
3. Platy microstructure: when elongated and horizontal aggregates are stacked together, they are usually separated by planar voids.
4. Lenticular microstructure: elongated but lenticular aggregates (plates) are stacked and isolated from each other by sub-horizontal, wavy, and/or zigzag planar voids.
5. Massive microstructure: an abundant solid phase in which it is impossible to isolate aggregates or peds. Some small voids can be observed, but in the picture, white spots are quartz grains and not voids.
6. Complex microstructure: a mixture of two or more microstructural types. In this case, the microstructure can be named using a combination of terms defining the various microstructures observed in the soil thin section.
7. Spheroidal microstructure, or onion skin microstructure (FitzPatrick 1993): a specific microstructure recognized in the nomenclature adopted by Stoops (2003). The shape is given by discontinuous and concentric curved planar voids developed in a groundmass, XPL view.
8. Vermicular microstructure: it resembles an intertwining mass of worm shapes. It is a complex network of dense complete infillings lined by continuous or discontinuous wormlike voids. Introduced by FitzPatrick (1993), it is a specific microstructure also recognized in the nomenclature used by Stoops (2003).

Chapter 3

Basic Components

File 22: Mineral and Organic Constituents

Mineral and organic constituents belong to the basic components observed in soil thin sections. They can appear, for instance, as large rock fragments, or single minerals as sand grains; they can constitute large areas of micromass formed by clay minerals or display parts of plant roots or leaf fragments, i.e. organic material. These constituents comprise the body of the soil itself, and in soil micromorphology, they belong to the groundmass, as well as the material constituting the pedofeatures (see "File 9"). Two types of basic components are recognized by Stoops (2003, 2021), those recognizable at the magnifications of the optical microscope and those which are not. Stoops (2003, 2021) pointed out the problem of the optical microscope resolution and the thickness of conventional thin sections. Indeed, it is preferable not to have a standard size limit between coarse and fine materials. Consequently, there are three main types of basic components: the coarse mineral constituents, the fine mineral phase, and the organic matter-related constituents.

After a description of the sorting and shape of coarse grains, five sections present the main rocks encountered in soils. During their weathering phase, rocks can free mineral grains in the soil environment, and this is illustrated in five other sections. Three sections of rocks show the large diversity of the micromass and the other constituents of the groundmass. Some minerals do not originate from rocks but from living organisms, such as plants, bacteria, fungi, and animals: they are known as "biominerals" and three sections report the most common of them. Organic matter plays a fundamental role in soils and it leaves many traces of its impact at the microscale: three sections describe its various characteristics. Finally, soils were the foundations of all civilizations: they often contain the traces of humankind, which are called anthropogenic features. These are illustrated by two sections.

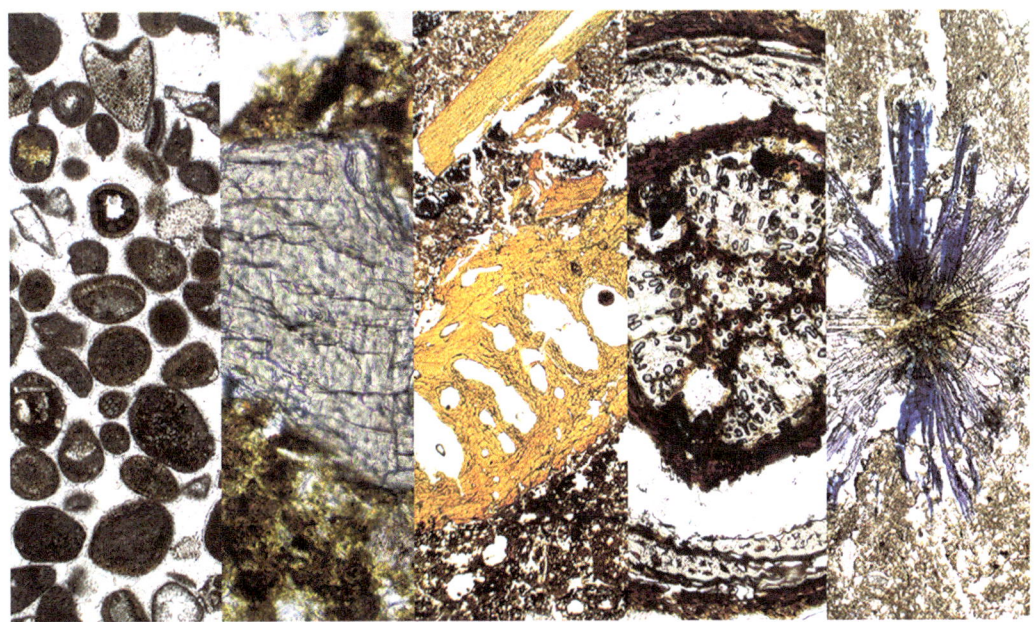

The diversity of constituents, from minerals to organic or anthropogenic features. All microphotographs are in PPL.

E. P. Verrecchia, L. Trombino, *A Visual Atlas for Soil Micromorphologists*, https://doi.org/10.1007/978-3-030-67806-7_3

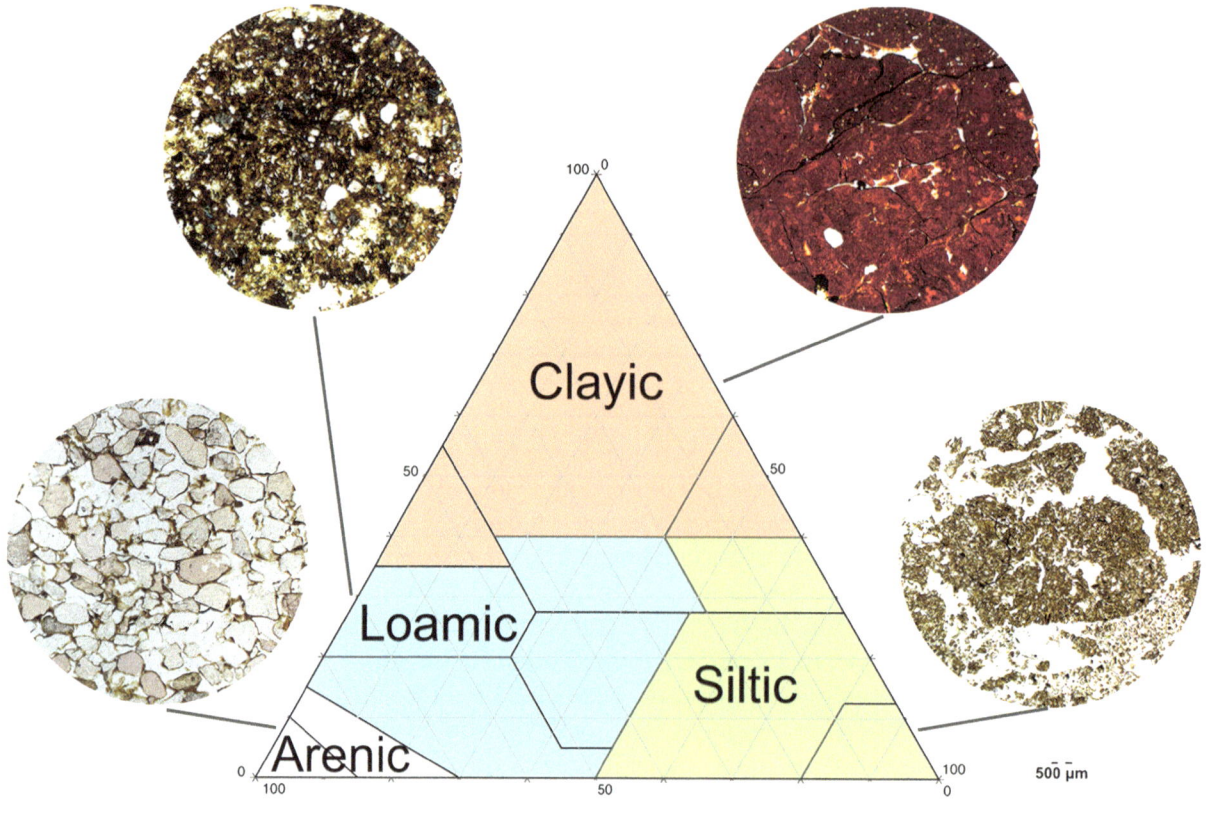

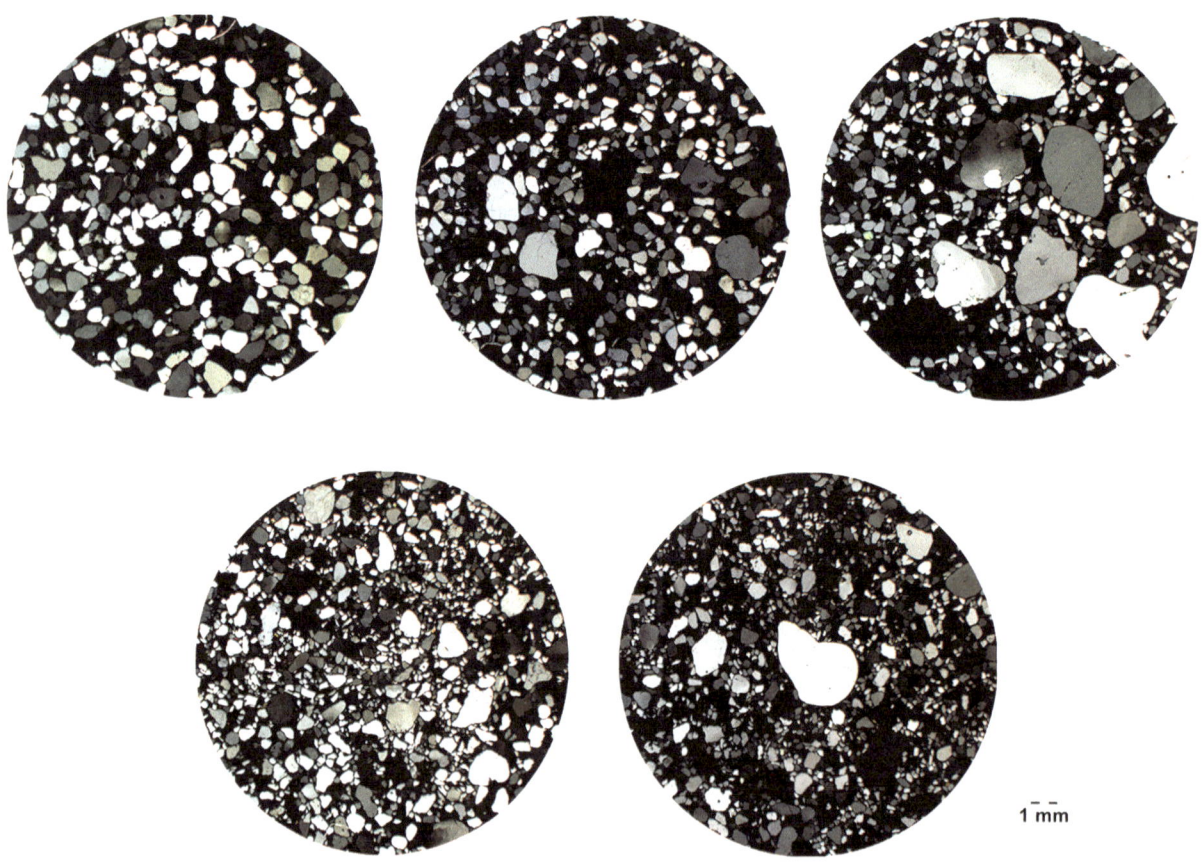

File 23: Particle Size and Sorting

The proportions of coarse and fine materials, according to their size, their degree of sorting, and their shape, constitute the fundamental parameters related to the soil texture in thin sections. All these terms are currently used in sedimentary petrography to describe terrigenous clastic sedimentary rocks; they are also used in soil science, as far as physical soil properties are concerned. Moreover, regarding particle sizes, it must be stressed that it is difficult to measure the sizes of constituents in thin sections, as they depend on the orientation of the object and its cross-cut plane in respect to the thin section surface.

Captions start with the ternary plot at the top. Round images below the ternary plot are listed from the upper left corner to the lower right corner.

1. Ternary plot of soil particle sizes and texture classes according to the World Reference Base for Soil Resources (IUSS and Working-Group-WRB 2014). Examples of texture microphotographs (in PPL) are shown for each type of texture, i.e. *clayic*, *siltic*, *arenic*, and *loamic*. These examples are illustrated by actual soil samples for which analyses of particle-size distributions have been performed. Clayic texture: soil from the Ligurian coast (northern Italy); siltic texture: soil from the Loess Plateau (China); loamic texture: soil from the northern Apennines (Italy); arenic texture: soil from the Central Sahara (Libya).

2.–6. After a grain-size distribution analysis of a primary material, the classes obtained have been selectively mixed together in order to reconstruct visual examples with suitable distributions of grain sizes. All views in XPL.

2. Only one size fraction of quartz sand is present, making the sorting perfect.

3. Only up to 10% of grain-size fractions other than the dominant quartz sand size fraction are present, resulting in a well sorted soil texture.

4. Fractions other than the dominant quartz sand size fraction represent 10–30% of the mineral grains, resulting in a moderately sorted soil texture in the thin section. At least two different fractions are obvious: fine and very coarse.

5. The well sorted quartz sand fraction is no longer the dominant one, resulting in poor sorting. No single size is clearly dominant.

6. A variety of quartz sand sizes is present in similar proportions. The sorting is very poor and the sediment is qualified as unsorted.

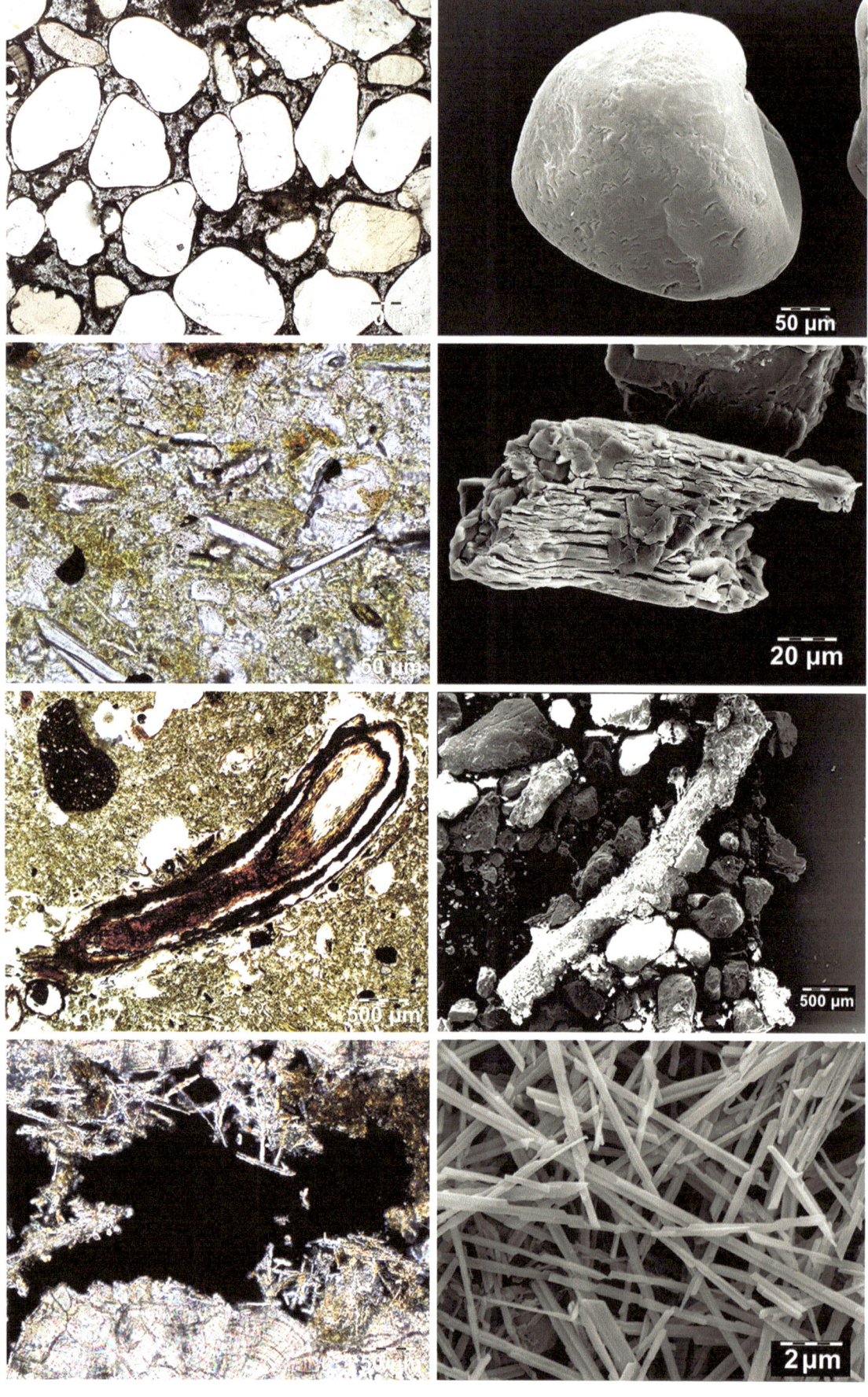

File 24: Shape of Grains: Equidimensionality

Equidimensionality refers to the way particle sizes are organized regarding the three perpendicular dimensions of space and how equal they are. However, particle shapes can only be described according to two dimensions in thin sections; therefore, the real three-dimensional morphology of a particle must be deduced or inferred very carefully, because it depends on the orientation of the object and its cross-cut plane in respect to the thin section surface.

Captions from upper left corner to lower right corner.

1. Equant quartz sand grains surrounded by thin layers of micrite and cemented by microsparite (Bizerte coast, Tunisia; XPL). The equant morphology is characterized by the same amplitude along the three axes.
2. Three-dimensional scanning electron microscope image of a single equant quartz sand grain of aeolian origin (Chobe Enclave, Botswana).
3. Planar mica grains dispersed in a yellowish grey micromass (northern Po Plain, Italy; PPL). In the planar morphology, one axis is much shorter than the two others, giving a flat shape to the object.
4. Three-dimensional scanning electron microscope image of a planar mica crystal of the same size as in 3. (Jura Mountains, Switzerland).
5. Prolate rootlet fragment inside a channel in a loamy soil (northern Apennines, Italy; XPL). Such morphologies are usually oblong, with one axis longer than the others.
6. Three-dimensional scanning electron microscope image of a prolate rootlet fragment in a coarse sandy soil (central Po Plain, Italy).
7. Acicular crystals of needle-fibre calcite, coating a channel in a silty soil (Loess Plateau, China; XPL). Like the calcite crystals shown in the photograph, the acicular shape refers to needle-like morphologies.
8. Three-dimensional scanning electron microscope image of acicular crystals of needle-fibre calcite (Jura Mountains, Switzerland).

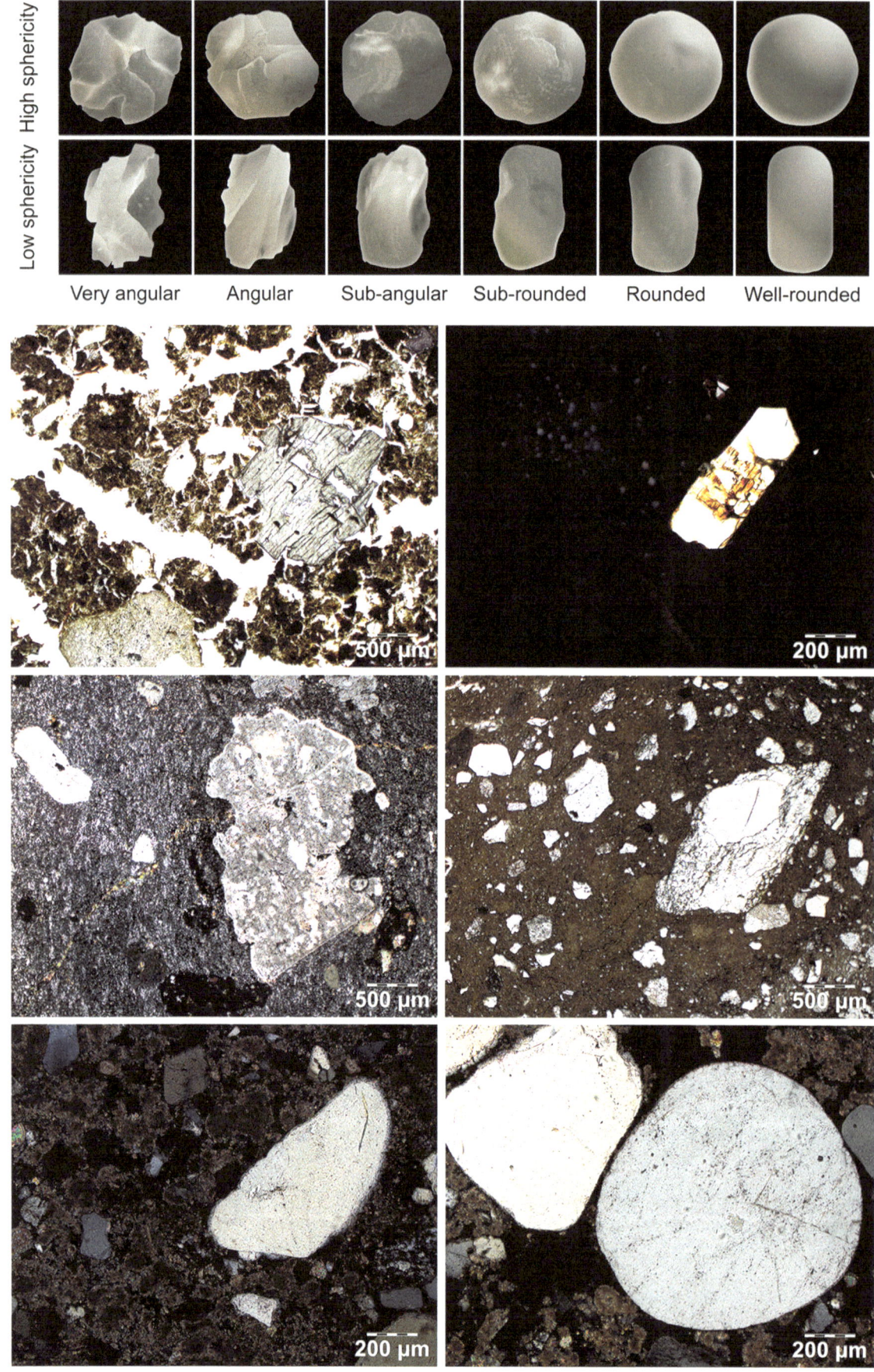

High sphericity / Low sphericity

Very angular   Angular   Sub-angular   Sub-rounded   Rounded   Well-rounded

File 25: Shape of Grains: Roundness and Sphericity

The roundness of a particle is determined by the sharpness of its edges and corners, independently of the shape of the particle itself. The sphericity of a particle is determined by its overall form, independently of the sharpness of its edges and corners. Both properties are commonly evaluated by means of visual estimation charts, even if today, image processing software can automatically generate parameters describing the roundness and sphericity of individual particles (see "File 77").

Captions start with a visual estimation chart at the top. Then, they are listed from upper left corner to lower right corner.

1. Visual estimation chart of both roundness and sphericity, redrawn from Powers (1953). Six classes of roundness and two classes of sphericity are provided. In the original work of Powers (1953), the particles were modelled from clay to make possible the addition of details like shape, sphericity, and roundness of particles. Many other charts have been drawn since, but whatever the type, they all are based on the same approach as Powers (1953); they usually add some intermediate steps in the variability of parameters. The roundness classes of Powers (1953) are based on a specific ratio $\rho = \frac{r}{R}$, where r is the radius of curvature of the largest inscribed circle and R is the radius of the smallest circumscribing circle. The ratio's values range from the "very angular" to the "well rounded" classes, as follows: 0.12 to 0.17, 0.17 to 0.25, 0.25 to 0.35, 0.35 to 0.49, 0.49 to 0.70, and 0.70 to 1. Based on Wadell's work (Wadell 1932), Krumbein (1941) proposed to estimate the sphericity by calculating $\Psi = \left(\frac{bc}{a^2}\right)^{1/3}$, where a, b, and c are the long, intermediate, and short axis dimensions (respectively) of the particle. Today, quantitative image processing automatically allows particles to be selected and shape parameters to be calculated (Heilbronner and Barrett 2014).
2. High sphericity, very angular: amphibole grain from an Andosol, tropical eastern Africa; PPL.
3. Low sphericity, angular: weathered feldspar grain in a dense dark micromass from an Andosol, tropical eastern Africa; XPL.
4. Low sphericity, subangular: micritic to microsparitic nodule from a Vertisol, northern Cameroon; XPL.
5. Low sphericity, sub-rounded: quartz grain in a carbonate micritic micromass of a Calcisol, Morocco; PPL.
6. Low sphericity, rounded: quartz grain associated with calcium carbonate pellets in a Calcisol, northern Tunisia; XPL.
7. High sphericity, well rounded: aeolian quartz grain observed in a Kastanozem, Botswana; XPL.

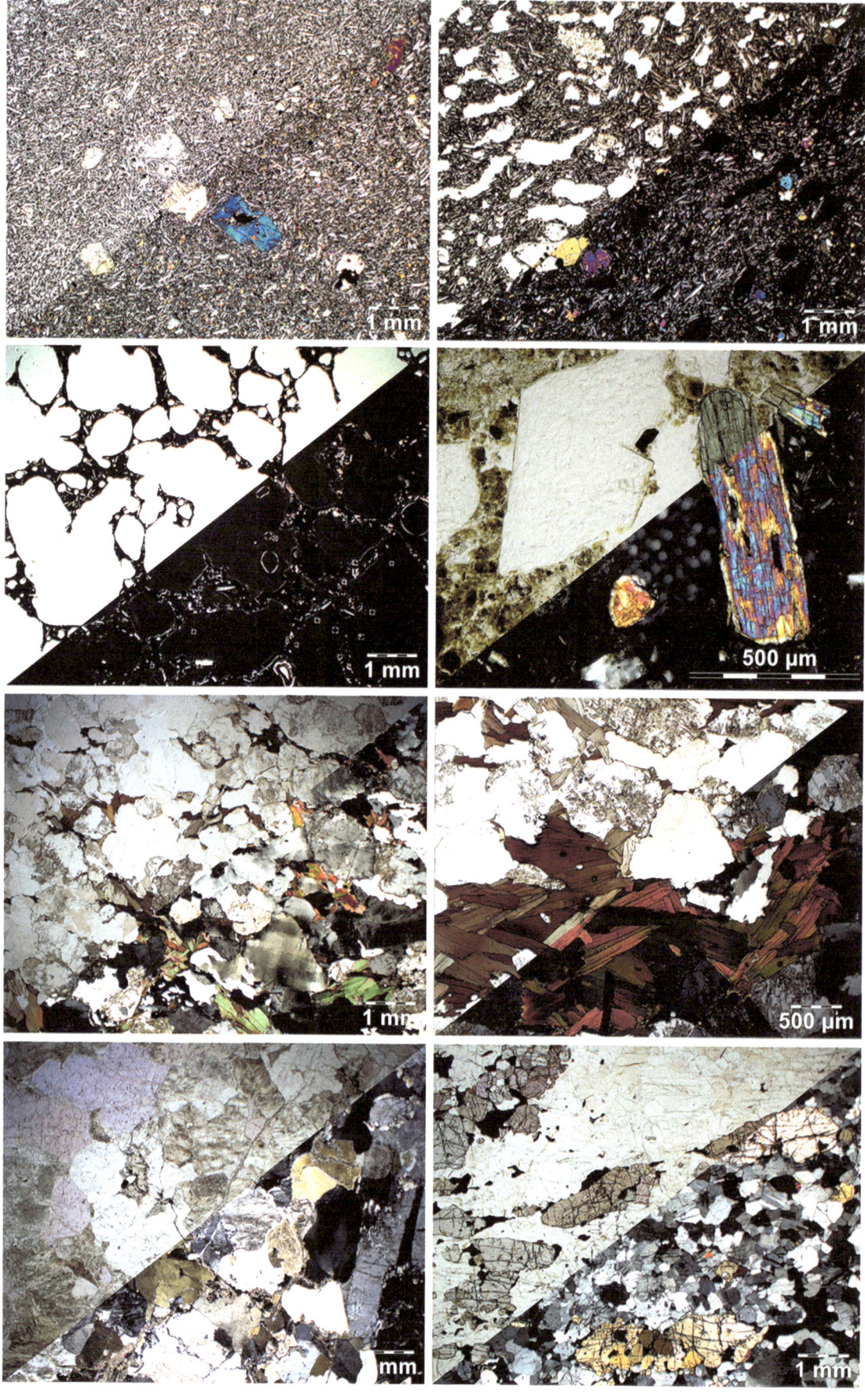

File 26: Basalt, Granite, and Gabbro

> It is very common to observe large pieces of fragmented rocks (lithoclasts) in soil thin sections. The recognition of their texture and nature is therefore important as they emphasize the role of soil parent material or can be the result of reworking of an observed soil horizon, indicating an allochtonous origin of the soil material. This section describes three common igneous rocks, i.e. basalt, granite, and gabbro.

Captions from upper left corner to lower right corner.

1. Basalts are dark volcanic igneous rocks with a microlitic texture, mainly composed of tiny crystals of plagioclases, forming the fine groundmass in the picture. Bright blue and orange crystals in XPL are phenocrysts of olivine minerals, and the whitish ones are calcic clinopyroxenes; Massif Central, France.

2. Vesicular basalt has the same texture and minerals as in 1., but with large voids due to the presence of gas. Such a microstructure makes this type of basalt more prone to weathering; Massif Central, France.

3. Basaltic pumice stone with large voids trapping air, separated by thin layers of basaltic groundmass with bubbles; Massif Central, France.

4. Volcanic ash deposited close to the eruption source. Large crystals are surrounded by weathered glass, an amorphous product constituting the micromass (extinct in XPL). The two large crystals are feldspar (in white in PPL) and amphibole (green in PPL and with birefringence colours typically up to middle second order in XPL); Massif Central, France.

5. Granite is a plutonic igneous rock with a granular texture. In this example from Forez, France, quartz is the dominant mineral (clear whitish crystals in PPL) associated with feldspars, such as microcline (microcline exhibits minute multiple twinning, which forms a cross-hatched pattern in XPL). Some micas are identified by their light-reddish orange or green colour in XPL.

6. Granite with large quartz grains (PPL and XPL); the crystals appearing clear and dusty in PPL, and grey in XPL, are feldspar. Brown (in PPL) and dark brown to reddish orange minerals are micas (biotite); Quebec, Canada.

7. Granite composed of quartz, orthoclase (a potassium feldspar), and plagioclases. Orthoclase appears as dusty crystals in PPL and grey in XPL. The bright yellow crystal is a mica. This granite underwent a degree of weathering evidenced by the fact that rare calcite crystals (dusty light-yellowish white in XPL) secondarily precipitated in some pores; Swiss Alps.

8. Gabbro is a common mafic igneous rock, generally greenish and dark-coloured. This sample from Ontario, Canada, contains many plagioclases (whitish in PPL and from white to grey and extinct in XPL) and minor amounts of olivine (light green in PPL and bright yellow in XPL).

File 27: Schist, Gneiss, and Amphibolite

> Some large pieces of lithoclasts in soil thin sections originate from metamorphic rocks. The recognition of their texture and nature is therefore important as they emphasize the role of soil parent material or can be the result of reworking of an observed soil horizon, indicating an allochtonous origin of the soil material. This section describes four common metamorphic silicate rocks, i.e. schist, gneiss, amphibolite, and greenschist.

Captions from upper left corner to lower right corner.

1. Typical schist texture showing sheet-like grains with a preferred orientation, i.e. NW–SE. The clear areas in PPL are mainly composed by quartz grains, whereas the fine layers, dark in PPL and pink to yellow in XPL, are oriented micas, i.e. biotite and muscovite; Italian Alps.
2. A mica-schist facies similar to 1. but with abundant thin layers of micas intertwined with thin quartz bands. Quartz is white, grey, and extinct in XPL, whereas mica appears in second order colours; Italian Alps.
3. A garnet-mica schist showing a garnet crystal in its centre (isotropic in XPL, i.e. black). The high-order coloured bands in XPL are comprised of muscovite mica with some brown crystals of biotite mica. The grey and white crystals in XPL are quartz with some feldspars; Italian Alps.
4. Example of a schist containing significant amounts of sericite, a fine-grained type of mica (high-order coloured small crystals in XPL). Remains of a plated structure of schistosity can be seen in the middle; Ontario, Canada.
5. An orthogneiss resembles a granite, as it results from the metamorphism of igneous rocks, with low amounts of mica (biotite in brown in XPL and muscovite with high-order coloured bands in XPL) and large crystals of quartz and feldspar (white, grey, and extinct in XPL); Swiss Alps.
6. A paragneiss designates a gneiss derived from a sedimentary rock. In this sample, micas are very abundant with bands of granoblastic interlocking crystals of quartz (white, grey, and extinct in XPL); Swiss Alps.
7. Metamorphic amphibolite formed by a dominant amphibole mineral, hornblende, associated with rare plagio-clases. Hornblende is green in PPL and displays bright interference colours of second order in XPL. Such rocks result from the metamorphism of igneous rocks (basalts or gabbros); Ontario, Canada.
8. Greenschist is a metamorphic rock usually produced by regional metamorphism. Its colour comes from the presence of large amounts of green minerals such as chlorite (green in PPL and second order interference colours) or epidote (small light green equant crystals in PPL). The schistosity is emphasized by the plate structure formed by the green minerals in PPL. Other minerals include feldspar and amphibole, as well as rare quartz; Ontario, Canada.

File 28: Quartzite and Marble

> Some large lithoclasts in soil thin sections originate from metamorphic rocks. The recognition of their texture and nature is therefore important as they emphasize the role of soil parent material or can be the result of reworking of an observed soil horizon, indicating an allochtonous origin of the soil material. This section describes two common rocks that result from the metamorphism of former sedimentary rocks, i.e. quartzite and marble.

Captions from upper left corner to lower right corner.

1. When a sandstone undergoes metamorphism, it can be transformed into a quartzite rock (example from Ontario, Canada). This picture taken in PPL does not show any particular pleochroism, all of the thin section being composed by colourless quartz.
2. Same view as in 1. in XPL. All the quartz crystals display either colours from the first order or are extinct. The minerals form a granoblastic structure typical of quartzite, i.e. the crystals have very irregular boundaries, and they are interlocking and appear as a mosaic. This quartzite is composed of quartz phenocrysts.
3. Same type of rock as in 1. in PPL. This quartzite from Ontario, Canada, is also composed solely of quartz grains.
4. Same view as 3. in XPL. The granoblastic structure is still present, but the size of the average quartz grain is much smaller than in 2., although some large crystals can be observed at the top of the photograph.
5. Marble is a metamorphic rock composed of recrystallized carbonate rocks made up of calcite or dolomite. In PPL, it is possible to see only the border of the phenocryst of calcite, the mineral being colourless in PPL; Quebec, Canada.
6. Same view as in 5. in XPL. The phenocrysts of sparite clearly appear as interlocking crystals of calcite with bright twin bands visible in extinct minerals. Calcite usually displays high-order colours of birefringence; Quebec, Canada.
7. Fine-grained marble, a metamorphic rock composed of recrystallized calcite from a primary sedimentary mud. There are only a few crystals of sparite, the rock fabric being mainly composed of fine microsparite. PPL, Quebec, Canada.
8. Same view as in 7. in XPL. The multiple and various high-order colours of birefringence denote the presence of numerous small crystals oriented in various directions; these directions can be mapped using electron backscatter diffraction (EBSD; see "File 77"). Some twinning zones can be observed in the larger crystals at the centre of the microphotograph; Quebec, Canada.

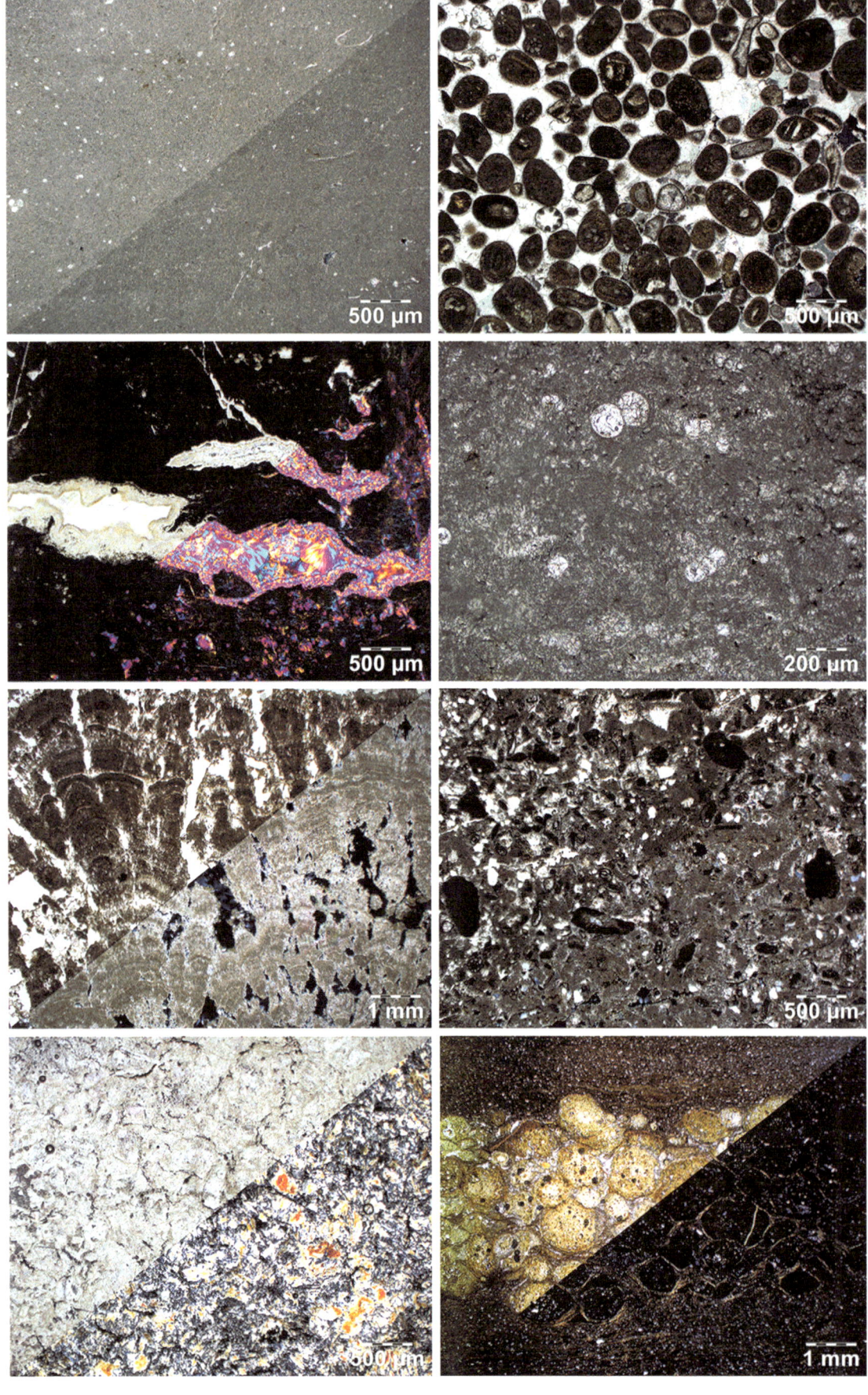

File 29: Calcium-Bearing Sedimentary Rocks

> Calcium is the fifth most abundant element on Earth. It is a major compound of numerous sedimentary rocks. Deciphering its origin remains a fundamental issue in soil science, as calcium is a crucial element in many pedogenic processes and biogeochemical pathways (Rowley et al. 2018). Although calcium is incorporated in many structural formulae of silicates, its supply from calcium-bearing sedimentary rocks largely prevails in soil development.

Captions from upper left corner to lower right corner.

1. Fine-grained limestone composed of micrite. Such limestones formed as mud and became hardened during diagenesis. When extremely fine, calcite has a greyish brown colour in PPL and high-order colours of birefringence in XPL; Jura Mountains, Switzerland.
2. Oolitic limestone with two generations of cements: an isotropic and early diagenetic rim of small and pointy crystals and large sparitic crystals infilling voids due to late diagenetic precipitation. Such limestones are observed in agitated marine environments where oolites (spherical grains composed of concentric layers) form. PPL; Jura Mountains, Switzerland.
3. Silicified limestone from the Aquitaine Basin, France. The micritic limestone underwent a late diagenetic process during which two siliceous cements precipitated: small crystals of chalcedony coat the pores, and large chalcedony spherulites infill the voids. The second order colours of the siliceous crystals in XPL are due to the use of a gypsum plate, which shifts the birefringence of minerals by one wavelength.
4. Senonian chalk sample composed of micrite and round foraminifera microfossils (*Globigerina sp.*). PPL; Galilee, Israel.
5. Meteogene travertine, according to Pentecost (2010)'s definition, formed by stacked irregular shrubs precipitated by *Schizothrix fasciculata* (Freytet and Verrecchia 1998) with some internal dark laminations, which can be seen in PPL and XPL; Aquitaine Basin, France.
6. Marl is a sedimentary rock in which a calcium carbonate phase is mixed with some clays and quartz grains. In XPL, some quartz grains appear in white, grey, or are extinct (Negev Desert, Israel).
7. Gypsum is a common deposit of evaporitic environments from the past. Gypsum can form large beds or invade cracks in rocks. Swiss Alps.
8. Lamina of calcium phosphate interstratified in an organic-rich marly mudstone from California, USA. The mudstone is dark brown in PPL, whereas phosphates appear as light yellow to brownish micro-nodules coated by clays. The calcium phosphate phase, with low birefringence, appears extinct in XPL, and clay coatings are emphasized by their light rims in XPL.

File 30: Sand and Sandstone

Sands are usually composed by loose detrital particles of rock fragments between 0.05 and 2 mm. Commonly, quartz grains predominantly comprise sands at the surface of continents. During diagenesis, sands undergo hardening by cementation, forming sandstones. Cements can have various mineralogical compositions, from silica to calcium carbonate, from iron oxyhydroxides to sulphates.

Captions from upper left corner to lower right corner.

1. Loose grains of medium to coarse sand, mainly quartz in origin, from an aeolian deposit. Grains are separated by packing voids, resulting from the simple arrangement of the loose particles. Italian Alps.
2. Very fine fluvial sand from an alluvium of the Rhone River, Switzerland. Layers of different grain sizes, emphasized by the sizes of the packing voids, separate the succession of short sedimentary events of various intensities.
3. Pure fine sandstone composed solely of very well-sorted quartz grains. A thin siliceous cement keeps the quartz grains attached; Fontainebleau Forest, France.
4. Pure medium sandstone made of well sorted quartz grains with rare micas. As in 3., the sandstone is hardened by a siliceous cement. Swiss Alps.
5. A calcareous sandstone made of quartz grains cemented by very large monocrystals of calcite. The large crystalline nature of the cement is emphasized in the XPL view in which three single sparitic crystals can be identified by their polarising colours: pinkish, blueish, and extinct. Such cements have generally two possible cementation origins: (a) in a phreatic environment or (b) during late diagenesis after burial. Sample from a cone-in-cone structure (Aassoumi et al. 1992), Morocco.
6. A calcareous sandstone made of well rounded and sorted quartz grains, coated with micrite and cemented by microsparite. Bizerte coast, Tunisia.
7. Ferruginous sandstone with heterometric quartz grains cemented by iron oxyhydroxide, which are dark brown in PPL and seem extinct in XPL. Aquitaine, France.
8. Well sorted quartz grains cemented by malachite forming a green sandstone. Malachite is a copper carbonate easily identified by its green colour in PPL. Negev Desert, Israel.

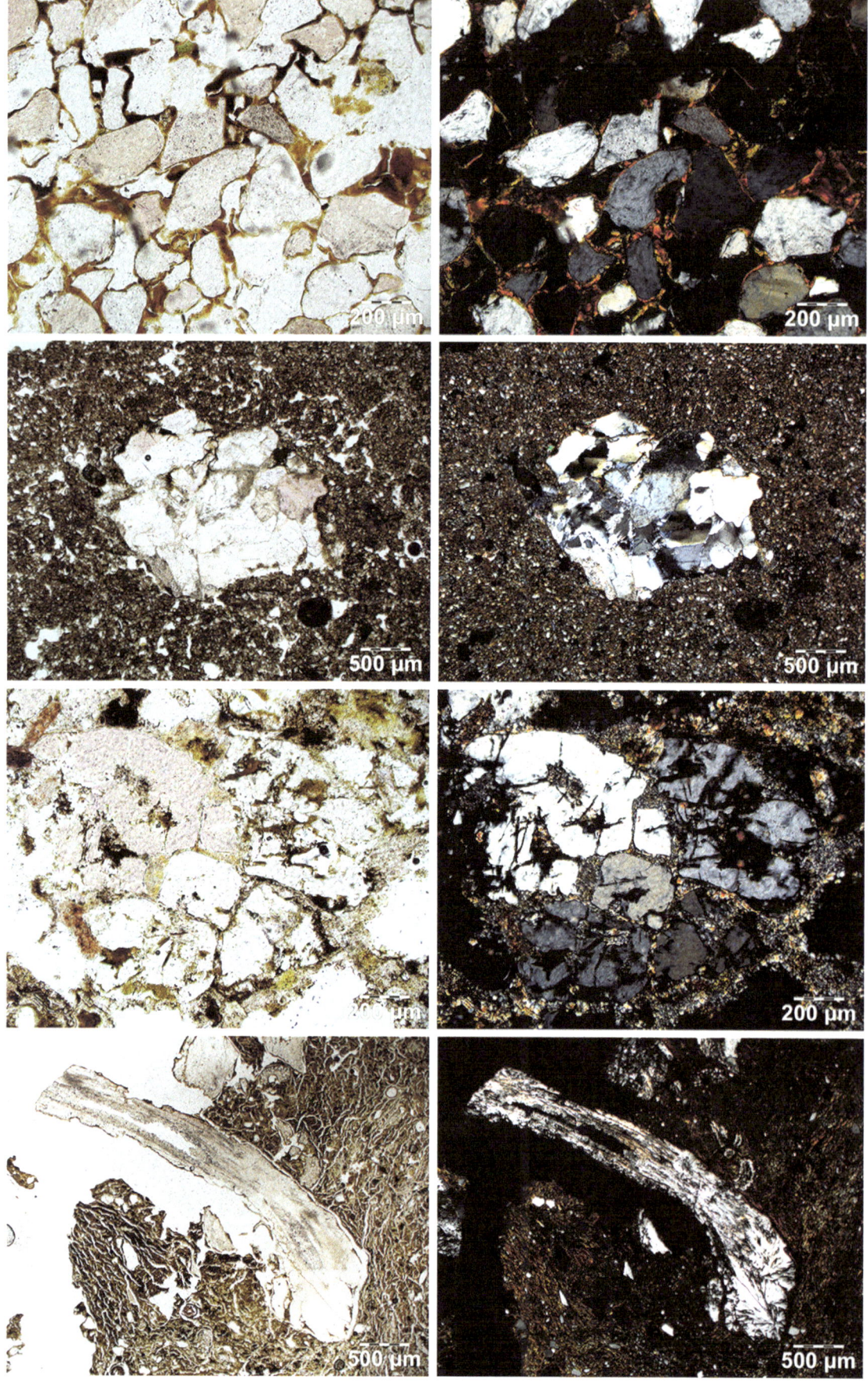

File 31: Mineral Grains in the Soil I: Quartz and Chalcedony

> Siliceous grains include quartz, its various polymorphs, and amorphous to cryptocrystalline siliceous minerals, for example chalcedony in chert. In volcanic material, siliceous grains can appear as cristobalite or tridymite.

Captions from upper left corner to lower right corner.

1. PPL view of individual and monocrystalline quartz grains in an arenic grain-size soil characterized by a chitogefuric c/f relative distribution with packing voids. Quartz grains are white to light beige, sub-rounded with a low sphericity (see "File 25"). The fine material is made of clays associated with iron oxyhydroxides. Paleosol, Libyan Sahara.
2. XPL view of the same field as in 1. Quartz grains appear in white, grey, or extinct (black), with weak birefringence to first order greys (see "File 8"). Grains are rimmed by a brownish to orange coating made of clays associated with iron oxyhydroxides.
3. Rock fragment composed of multiple quartz grains from a metamorphic rock (see "File 28"). The quartz grains appear colourless and non-pleochroic, i.e. they do not show any variation in colour when rotating the circular sample stage. Crystals have various shapes, without any clear cleavage, and display a very low relief. The fragment floats in a silty–clayey groundmass. Cambisol, Jura Mountains, Switzerland.
4. Same view as in 3. in XPL. Quartz in the fragment displays various first order greys, from white to black (extinct). However, the extinction of some individual quartz crystals are not uniform (shadowy), a common feature of quartz in deformed rocks. A closer view reveals that some quartz grains have lobate and interfingering boundaries, and others show dark and light veinlets penetrating into the crystal, emphasizing twin lamellae.
5. Aggregate of clear quartz grains with small and dark inclusions (PPL view). Jura Mountains, Switzerland.
6. Same view as in 5. in XPL. The centre of the quartz grains displays typical features such as large open cracks, embayments, and what were probably melt inclusions that were later infilled with weathering products, such as clays and amorphous silica (likely chalcedony). Such quartz is probably of volcanic origin.
7. View in PPL of a colourless flint flake in a hockey stick shape embedded in a loamy groundmass. Jura Mountains, Switzerland.
8. Same view as in 7. in XPL. Chalcedony is characterized by very low grey to whitish colours. The shaft of the "hockey stick" is made of length-slow chalcedony, whereas the blade shows spherulites of length-fast and zebraic chalcedony.

File 32: Mineral Grains in the Soil II: Feldspar and Mica

Feldspars are tectosilicates commonly found in igneous and metamorphic rocks (see "File 26"). They are characterized by two mineral families: the K-feldspars and the plagioclases, which form a continuous solid solution with various amounts of sodium and calcium, from albite (a sodium feldspar) to anorthite (a calcium feldspar). During weathering, calcium, sodium, and potassium ions are freed in the soil solution, and feldspar partially transforms into clays and/or oxyhydroxides. Micas are phyllosilicates frequently associated with granite and granodiorite but also with metamorphic rocks, such as schist and gneiss (see "File 26" and "File 27"). Muscovite is frequently observed as grains in soils, whereas biotite is rarer, as it is easily weathered and transformed into clay. This difference between the two micas is related to their structural chemical formula, the aluminium in muscovite being much more refractory to weathering than the magnesium and iron in biotite.

Captions from upper left corner to lower right corner.

1. Translucent grains in a fine brownish groundmass. It is difficult to clearly identify the mineralogical nature of the grain, although some faint parallel lines can be made out on the right hand side of the central grain. PPL view, Vertisol, Cameroon.
2. Same view as in 1. in XPL. The central grain is a plagioclase characterized by its polysynthetic twins of alternating dark and light lamellae. The two other grains are quartz. The groundmass is formed of clay minerals, fine quartz grains, and a small flake of light blue muscovite.
3. Large translucent grain with thin parallel and alternating white and extremely pale pink lamellae. PPL view, Fluvisol, Cameroon.
4. Same view as in 3. in XPL. Crystal of microcline showing its characteristic twins closely interwoven forming a typical cross-hatched ("tartan") pattern.
5. It is extremely difficult to find a fresh crystal of biotite in soils, due to their fast weathering. This brownish platelet preserves some characteristics of biotite (see "File 26"). PPL view, Fluvisol, Cameroon.
6. Same view as in 5. in XPL. The crystal of biotite is surrounded by quartz (from white to grey, or extinct), clays, and oxyhydroxides and still displays a greenish moiré colour due to exfoliation combined with formation of secondary minerals (probably vermiculite).
7. Translucent crystal of muscovite formed by visible thin sheets with a relatively high relief, showing cleavage. PPL view, Cambisol, Switzerland.
8. Same view as in 7. in XPL. The crystal appears in blue and pink bright birefringence colours typically up to middle third order. The alternating bright blue and pink colours underline the sheet structure of such crystals.

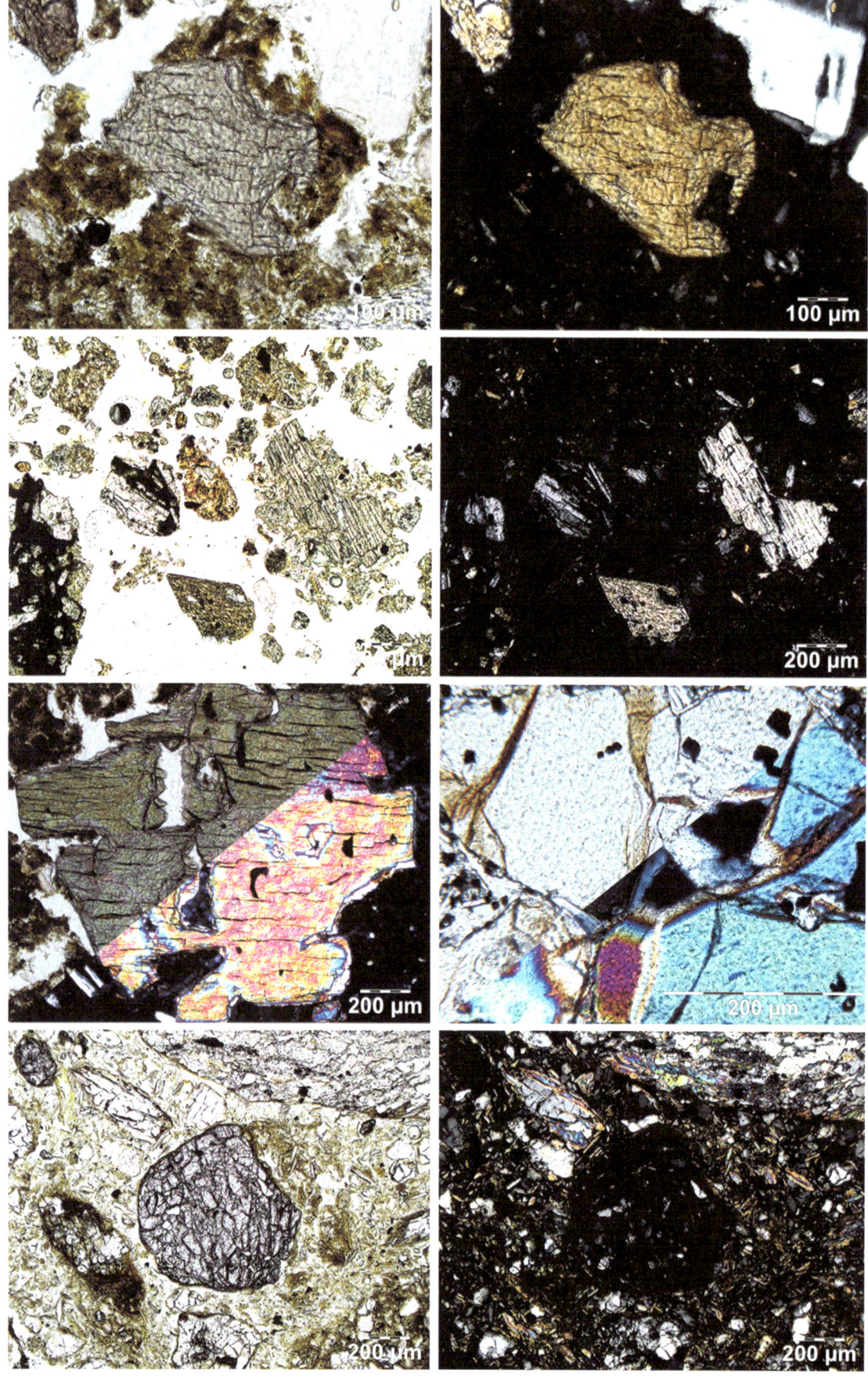

File 33: Mineral Grains in the Soil III: Inosilicates and Nesosilicates

Inosilicates are formed of interlocking chains of silicate tetrahedra and include two main groups of minerals: pyroxenes, as single chain silicates, and amphiboles, as double chain silicates. They are commonly igneous rock-forming minerals but can also be associated with high-temperature metamorphic rocks. In nesosilicates, the silicate tetrahedra are isolated and bound to each other by ionic bounds, making their structure particularly dense, providing them some resistance to weathering. Olivines (minerals found in high-temperature igneous rocks) and garnets (minerals prevalent in metamorphic rocks) form two common members of nesosilicates. Both of these rock-forming silicate groups, inosilicates and nesosilicates, can be observed in soils as residual mineral grains in the coarse fraction.

Captions from upper left corner to lower right corner.

1. Isolated grain of pyroxene, pale coloured with a subtle greenish pleochroism. The crystal is characterized by a high relief with two sets of cleavages at 90°. The pyroxene grain is surrounded by a brown, cloudy, and undifferentiated micromass. PPL view, Andosol, Central Africa.
2. Same view as in 1. in XPL. The colour of the crystal changes to light brown with the cleavage remaining perfectly visible. The micromass is totally extinct and corresponds to cryptocrystalline to amorphous phases.
3. Fragments of amphibole crystals showing brownish to pale yellowish green pleochroic colours in PPL. The shape of the pointy and half-diamond crystal (bottom part of the micrograph) is due to the 60° and 120° cleavage angles of amphiboles, emphasizing potential weakness of the crystal lattice during its weathering in soils. Andosol, Central Africa.
4. Same view as in 3. in XPL. Amphibole crystals show interference colours in the upper first or lower second order. The single orientation of cleavage in the upper crystal is due to its orientation. Other crystals in dark and light grey are feldspars.
5. Amphiboles sometimes display second order interference colours, here bright yellow to pink with shades of blue. Andosol, Central Africa.
6. Olivine with weathering products in cracks (light brown in PPL) formed by iddingsite. Olivine, usually colourless in PPL, displays typical second order interference colours in XPL (here bright blue). Andosol, Massif Central, France.
7. Garnets are colourless minerals with a high to extreme relief, forming equant crystals, without any observable cleavage. PPL view, Cryosol, Lombardy (Italy).
8. Same view as in 7. in XPL. As an optically isotropic mineral, garnet remains extinct in XPL. The micromass is formed of small quartz and muscovite crystals due to the weathering of metamorphic pebbles.

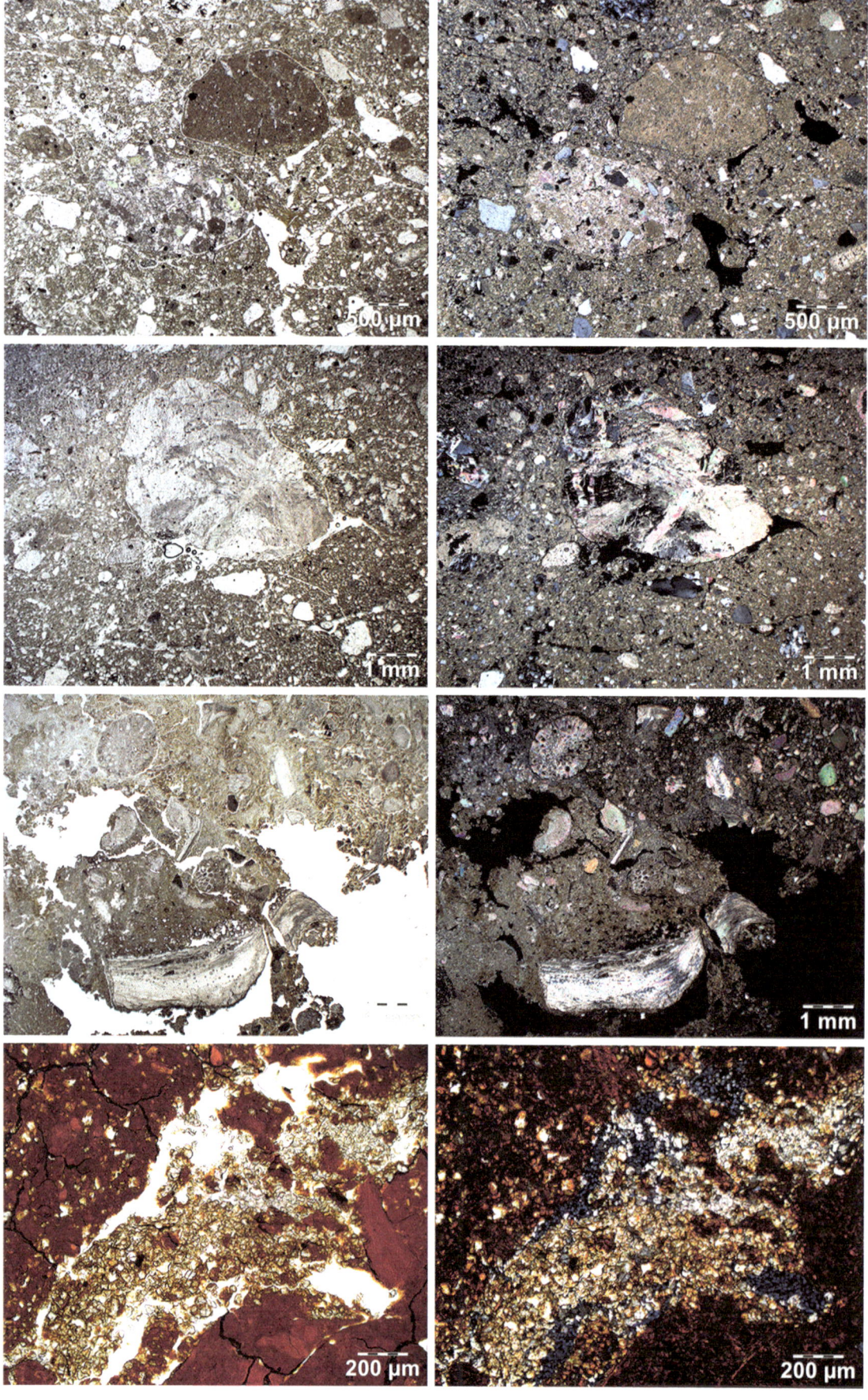

File 34: Mineral Grains in the Soil IV: Carbonates

Carbonate minerals are common features of soils. They can be either inherited from the carbonate parent material of the soil (see "File 29" for some examples) or precipitated as secondary pedogenic features (see e.g. "File 72"). In this section, carbonate grains found in soils are inherited from the bedrock. These elements in the soil coarse fraction constitute lithoclasts and not pedofeatures. Carbonate lithoclasts in soils are usually inherited from three main types of rocks: limestones, marls, and dolomites.

Captions from upper left corner to lower right corner.

1. Two large limestone lithoclasts integrated inside a soil rich in detrital carbonate grains. These lithoclasts are formed by inherited fragments of micritic (brown fragment) and microsparitic limestones. The brownish micromass is a mixture of micrite, clays, and very small quartz grains. PPL view, Calcaric Cambisol, Jura Mountains, Switzerland.
2. Same view as in 1. in XPL. The microsparitic nature of the bottommost fragment is emphasized in XPL by the high-order interference colours of calcite. Grey-coloured grains are quartz. Calcaric Cambisol, Jura Mountains, Switzerland.
3. Inherited sparitic detrital grain in the same type of soil as in microphotograph 1. Sparitic crystals are colourless with a low relief. PPL view, Calcaric Cambisol, Jura Mountains, Switzerland.
4. Same view as in 3. in XPL. The sparitic nature of this fragment is emphasized in XPL by the high-order interference colours of calcite. This large lithoclast is surrounded by small calcite and quartz grains in a brownish micromass. Calcaric Cambisol, Jura Mountains, Switzerland.
5. Fragments of an inherited and fossiliferous limestone with detrital grains of Bryozoans (e.g. top) and shells (bottom; probably an oyster shell) in a calcareous micromass. PPL view, Calcaric Cambisol, Jura Mountains, Switzerland.
6. Same view as in 5. in XPL. The calcium carbonate and partially fibrous nature of the bioclasts (organism fragments) are indicated in XPL by the high-order interference colours, as well as the interwoven and rolling extinctions. Calcaric Cambisol, Jura Mountains, Switzerland.
7. Inherited dolomitic crystals in a red clayey and iron oxyhydroxide-rich micromass. PPL view, Paleosol, Liguria, Italy.
8. Same view as in 5. in XPL. Dolomite crystals display their greyish interference colours although these colours are partially altered by the presence of red clayey and iron oxyhydroxide-rich thin coatings. Paleosol, Liguria, Italy.

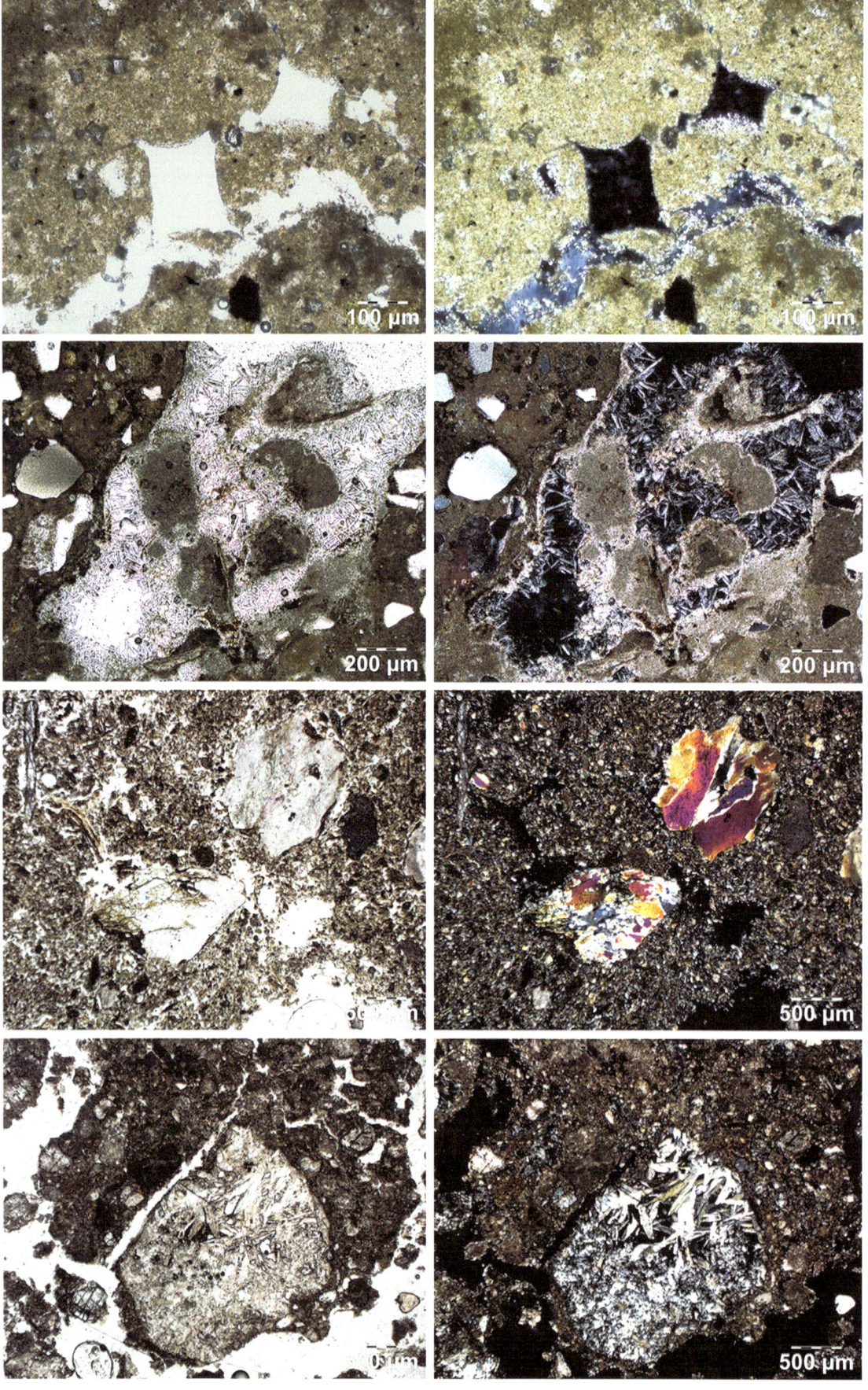

File 35: Mineral Grains in the Soil V: Chlorides and Sulphates

Although many inherited chlorides and sulphates could have been easily dissolved during weathering, some soils can preserve their clasts or imprints. They must not be interpreted in terms of pedogenic processes when they are inherited from the bedrock. It is why their recognition as lithoclasts is paramount. This plate shows examples of a chloride (halite) and three different types of sulphate: barite, anhydrite, and gypsum. In this plate, all these minerals are inherited from the bedrock.

Captions from upper left corner to lower right corner.

1. Crystals of halite in a carbonate and clayey micromass. They are recognizable by their square or rectangular shape due to their isometric crystallographic system (cubic form). As such, they appear white in PPL with a very low relief. Dead Sea, Israel.
2. Same view as in 1. in XPL. Halite is isotropic, i.e. the crystal does not display any interference colour and stays extinct in XPL. Dead Sea, Israel.
3. Colourless crystals of barite (a barium sulphate), showing a high relief, and associated with calcium carbonate (micritic) lithoclasts in a large pore of a palustrine limestone; PPL view, Chobe Enclave, Botswana.
4. Same view as in 3. in XPL. Crystals of barite are fibrous to columnar, up to first order pale yellow. Palustrine limestone in potential hydrothermal settings. Chobe Enclave, Botswana.
5. Lithoclasts of calcium sulphate anhydrite in a loessic groundmass. These mineral fragments are directly inherited from the bedrock and have been incorporated in the upper soil layer by bioturbation. These granular clusters of anhydrite remain colourless in PPL. Soils developed on Triassic gypsic sedimentary rocks. Saint-Léonard, Valais, Switzerland.
6. Same view as in 5. in XPL. The two large clusters of anhydrite display bright second and third order interference colours, mostly purple, red, and yellow, but green is also common. In the lower lithoclast, some first order grey and white interference colours indicate the presence of gypsum crystals (see below).
7. Lithoclast of colourless gypsum crystals in a clayey and carbonate micromass. The low-relief crystals form various habits, from elongated to granular shapes. PPL view, soils developed on Triassic gypsic sedimentary rocks. PPL view, Saint-Léonard, Valais, Switzerland.
8. Same view as in 7. in XPL. Gypsum is characterized by typical first order grey and white interference colours. The lenticular to acicular shapes of crystals are emphasized in the upper part of the cluster, whereas the lower part displays a saccharoid (granular) habit.

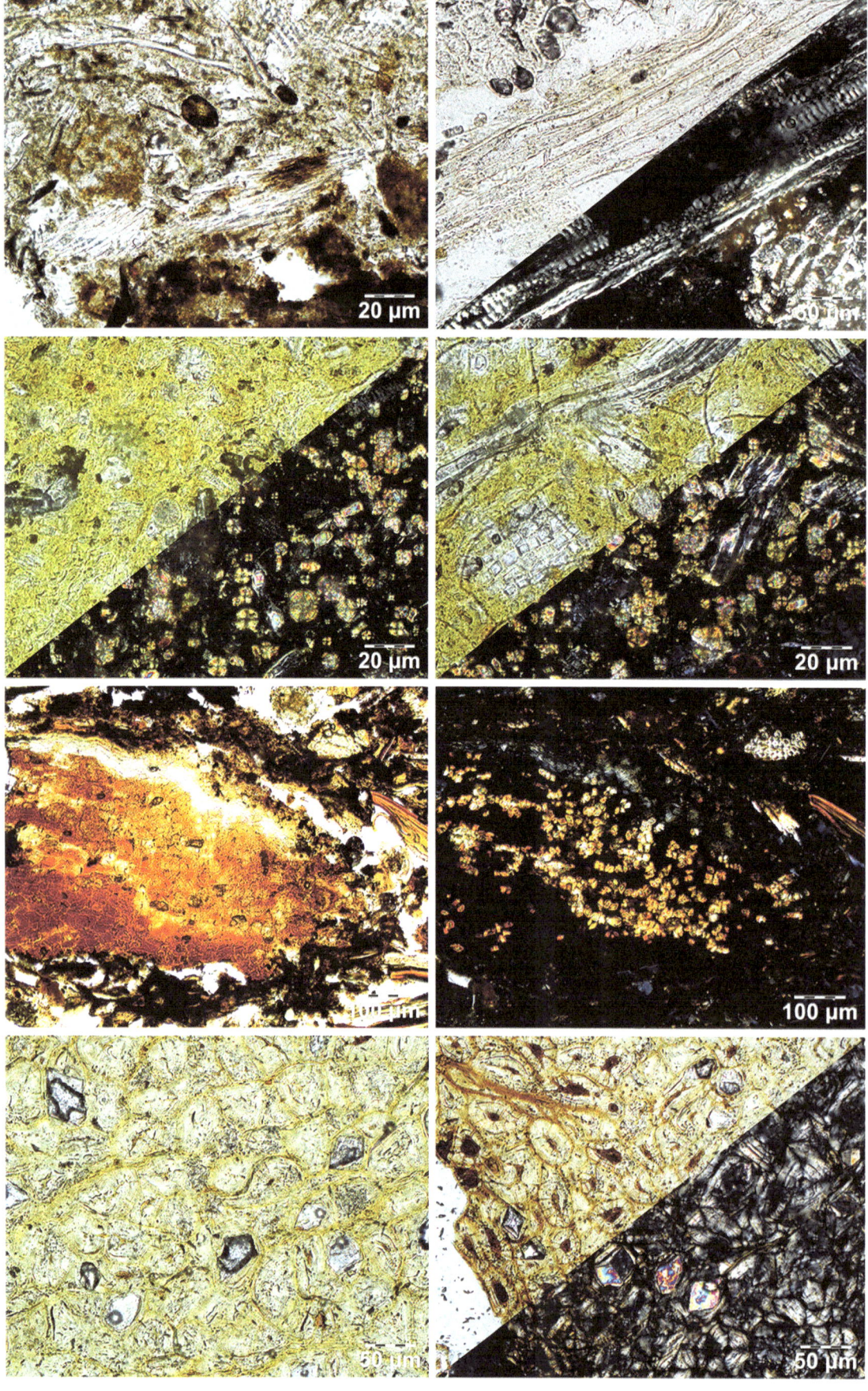

File 36: Biominerals I

Stoops (2003, 2021) defines biominerals as inorganic residues of biological origin. They include fragments of internal or external skeletons of animals, as well as direct or indirect mineral products of organism metabolism—for more details, see Mann (2001) and Skinner and Ehrlich (2017). Phytoliths are produced in specific plant cells, in intercellular spaces, or associated with cell walls Stoops (2003). Calcium oxalate druses are aggregations of crystals with a roughly overall spherical shape (Baran and Monje 2008), and they also originate from plants. Other forms of oxalate crystals exist, such as equant or needle-shaped morphologies, but both can be associated with plants and fungal filaments (Verrecchia et al. 1993). These minerals indicate the presence of organisms, even if the organic matter has been totally decayed.

Captions from upper left corner to lower right corner.

1. Clusters of opal phytoliths (transparent and elongated bodies) associated with plant residues (in brown). These phytoliths have been disarticulated and reworked during pedoturbation. PPL view, Libyan Sahara, archaeological deposit.

2. Assemblage of elongated opal phytoliths, transparent in PPL and with a low birefringence to extinct in XPL. The arrangement of the crystals precipitated inside the plant tissues is well preserved. Libyan Sahara, archaeological deposit.

3. Calcite spherulites associated with some oxalate druses in a dung deposit. In XPL, spherulites are characterized by a black cross, emphasizing their fibro-radial nature. Druses have a more radial structure with irregularly shaped borders. In PPL, these biominerals are difficult to observe.

4. Assemblage of essentially oxalate druse crystals with a radial structure and irregularly shaped borders in XPL. Same source as in 3.

5. Plant fragment in a dung deposit from the Sahara Desert. Part of the plant organic matter is decomposing, i.e. it appears as a dark colour. In the central part of the microphotograph, some transparent crystals are clustered, but difficult to identify. PPL view.

6. Same view as in 5. in XPL. Note the presence of equant crystals of calcium oxalate inside the remains of plant tissues.

7. Monoclinic oxalate crystals formed in specific cells of a tropical wood (iroko tree, *Millica excelsa*). Cellulose appears in light yellow and corresponds to the cambium. PPL view, Ivory Coast.

8. Slightly decayed iroko tree fragment, including oxalate crystals. In XPL, cellulose is characterized by a very high birefringence, whereas oxalate crystals display second order colours.

File 37: Biominerals II

Stoops (2003, 2021) defines biominerals as inorganic residues of biological origin. They include fragments of internal or external skeletons of animals, as well as direct or indirect mineral products of organism metabolism—for more details, see Mann (2001) and Skinner and Ehrlich (2017). This section mainly refers to three types of spherulites: (1) calcitic faecal spherulites result from animal digestive processes; they are more abundant in herbivore dung (Durand et al. 2018); (2) another type of calcitic and radial spherulites are found in desert laminar crusts, likely as a by-product of photosynthetic activity by cyanobacteria (Verrecchia et al. 1995); and (3) in bird and reptile droppings, it is possible to find spherulites, composed of uric acid (Canti 1998). These minerals indicate the presence of former biological activity, even if organic matter is not directly observed.

Captions from upper left corner to lower right corner.

1.–2. Faecal calcitic spherulites dispersed in a groundmass of dung in an archaelogical deposit, Libyan Sahara. In PPL (1), the crystals are almost invisible, whereas in XPL (2), they form an extinction cross due to their radial structure.

3.–4. Faecal calcitic spherulites, densely clustered, in archaeological dung, Libyan Sahara. Note the presence of a quartz grain at the bottom centre of the microphotograph.

5.–6. Uric acid spherulites dispersed in an organic groundmass found on a cave wall bed, Libya, Central Sahara. In PPL (left), they are not visible in the phosphatic groundmass. In XPL, they show a black extinction cross or a pseudo-isogyre resembling a biaxial interference figure (Canti 1998), visible in the centre of the microphotograph.

7.–8. Two examples of calcitic spherulites observed in laminar crust thin sections. Left: dense cluster of radial spherulites in a Pleistocene laminar crust from the Shephelah, Israel. Right: a thick layer of spherulites with a perfectly round fibro-radial crystal in the centre. Pleistocene laminar crust, southern Morocco. For their genesis, see details in Verrecchia et al. (1995).

File 38: Biominerals III

Stoops (2003) defines biominerals as inorganic residues of biological origin. They include fragments of internal or external skeletons of animals, as well as direct or indirect mineral products of organism metabolism—for more details, see Mann (2001) and Skinner and Ehrlich (2017). Mollusc shells are genetically induced biominerals (Mann 2001). The identification of the molluscs as originating from marine or continental environments is a clue to the nature of the soil parent material. Egg shells and bones also belong to genetically induced biominerals (Mann 2001): both of them are common features of archaeological sites (Durand et al. 2018). Finally, some clear and small crystalline aggregates are often found in temperate soils. In thin section, they appear as calcitic spheroids. They are usually formed by earthworms as metabolic by-products in their casts (Durand et al. 2018; Becze-Deak et al. 1997).

Captions from upper left corner to lower right corner.

1. Mollusc shell fragment showing a cross-section of the two layers composing the structure: the prismatic layer at the top and the lamellar layer at the bottom. Shell fragments can be made of either aragonite or calcite. Physical weathering of the prismatic layer can lead to the formation of calcite needle-fibres in soils (Villagran and Poch 2014). Archaeological site, Syria.

2. Cross-section of a shell with well-preserved chambers inside a micritic groundmass. This example is a freshwater snail (probably from the Planorbidae family). Holocene carbonate swamp deposits, Syria.

3.–4. Fragments of bird egg shells: the shell is formed by columnar calcite crystals (the "palisade layer"), oriented perpendicular to the surface of the egg, often with a narrow fan-like fabric.

5.–6. Fragment of a bone showing low interference colours (first order grey). This colour can be due to the remains of collagen, and not only apatite, which forms the mineral part of bones. Haversian canals (as channels) are frequently preserved (in 6. for example) and surrounded by fibres of apatite. Bone fragments are generally coloured from pale yellow to yellow or yellowish brown in PPL.

7.–8. Two examples of earthworm biospheroids from Cambisols, Switzerland. Calcitic biospheroids have mostly ellipsoidal shapes with a sharp and smoothly undulating or rough boundary, depending on the earthworm species. Their internal fabric resembles geodes, composed of drusic centripetal calcite crystals, the core being occupied by smaller crystals. Their density in thin section can vary greatly from one or two, to sometimes more than twenty.

File 39: Anthropogenic Features I

Many types of features in soils can be directly related to anthropogenic activities. In archaeological sites, their presence is obvious, but they can also be encountered in "natural" soils, providing evidence of the incidental presence of humankind (Nicosia and Stoops 2017). Pottery fragments are commonly found in archaeological sites. Chunks of brick can be present in more recent anthropogenic soils. Moreover, not only can artefacts be found in soils, but also secondary chemical deposits, such as amorphous phosphate or vivianite crystals. These deposits are frequently associated to reducing environments enriched in organic matter as, for instance, cesspits and latrines.

Captions from upper left corner to lower right corner.

1.–2. Ceramic shards from a Bronze Age archaeological site, northern Italy. It is mainly composed of a fired-clay matrix with some inclusion of coarse material and voids. The nature and content of compounds have varied over time and space and conditions of firing, but common traits indicate a heating effect on the clays, which can change the colour, the amount, and shape of planar voids. In 1., these changes can be seen as two yellow external layers bordering a brown central part. In 2., this change in colour is reversed.

3. Fragment of a raw (i.e. not fired) mud brick from a Bronze Age site, central Syria. It is composed of a clay matrix with some inclusions of coarse mineral grains and plant residues.

4. Fragment of a fired mud brick from a Medieval archaeological site from northern Italy. It is composed of a clay matrix with inclusions of coarse mineral grains with planar voids.

5. Fibro-radial cluster of vivianite crystals in a loamy groundmass. Same site as in 4. Its blue colour in PPL is diagnostic of this mineral. In XPL, vivianite displays a birefringence in the first order colours with a well-expressed pleochroism.

6. Cluster of vivianite crystals in elongated shapes of various sizes from a Medieval archaeological site from northern Italy. The blue colour of the crystal in PPL is diagnostic of this mineral.

7. A mix of amorphous phosphate and vivianite (blue in PPL) in a phosphatized clay and silty groundmass, from a Medieval archaeological site from northern Italy. In XPL, the b-fabric is undifferentiated and related to the impregnation of phosphate into the groundmass.

8. Same as in 5, except for the absence of vivianite and the coarser groundmass.

File 40: Anthropogenic Features II

Many types of features in soils can be directly related to anthropogenic activities (Nicosia and Stoops 2017). At archaeological sites, their presence is obvious, but they can also be encountered in "natural" soils. Indeed, a fire can be triggered by nature (e.g. summer forest fires or lightning strikes) or by humans. In this section, the features shown are related to anthropogenic activities induced by fire, i.e. charcoal, ashes, heated bones, and heated rock fragments.

Captions from upper left corner to lower right corner.

1.–2. Fragments of wood charcoals, opaque in PPL (as well as in XPL, not shown) with very fine (in 1.) and coarse (in 2.) porosity patterns, respectively. These patterns can be used by botanists to identify the type of wood. Moreover, the microstructure is fine granular and related to biological activity. Paleosol, northern Apennines, Italy.

3.–4. Ash deposits from an archaeological site in the central Sahara, Libya. Ashes, grey in PPL, are constituted by microcrystalline aggregates of calcium carbonate, clearly observed in XPL. The groundmasses of these examples also include numerous coarse grains of quartz and some elongated charred plant remains (in brown or black).

5.–6. Heated bone fragments found in a firepit with burned pebbles from a Neolithic site, northern Italy. In 5., the bone colour is dull orange-reddish brown in PPL and bright orange-reddish in XPL. This colour could correspond to a heating temperature of 400 °C (Villagran et al. 2017). In 6., the heated zone is limited to the upper left side where the colour is brownish, while the bone displays a light to medium yellow colour in PPL, corresponding to a lower heating temperature (Villagran et al. 2017).

7.–8. Heated flint/chert shards found in a firepit with burned pebbles from a Neolithic site, northern Italy. They exhibit changing colours due to oxidation. In addition, these shards are surrounded by a clayey micromass.

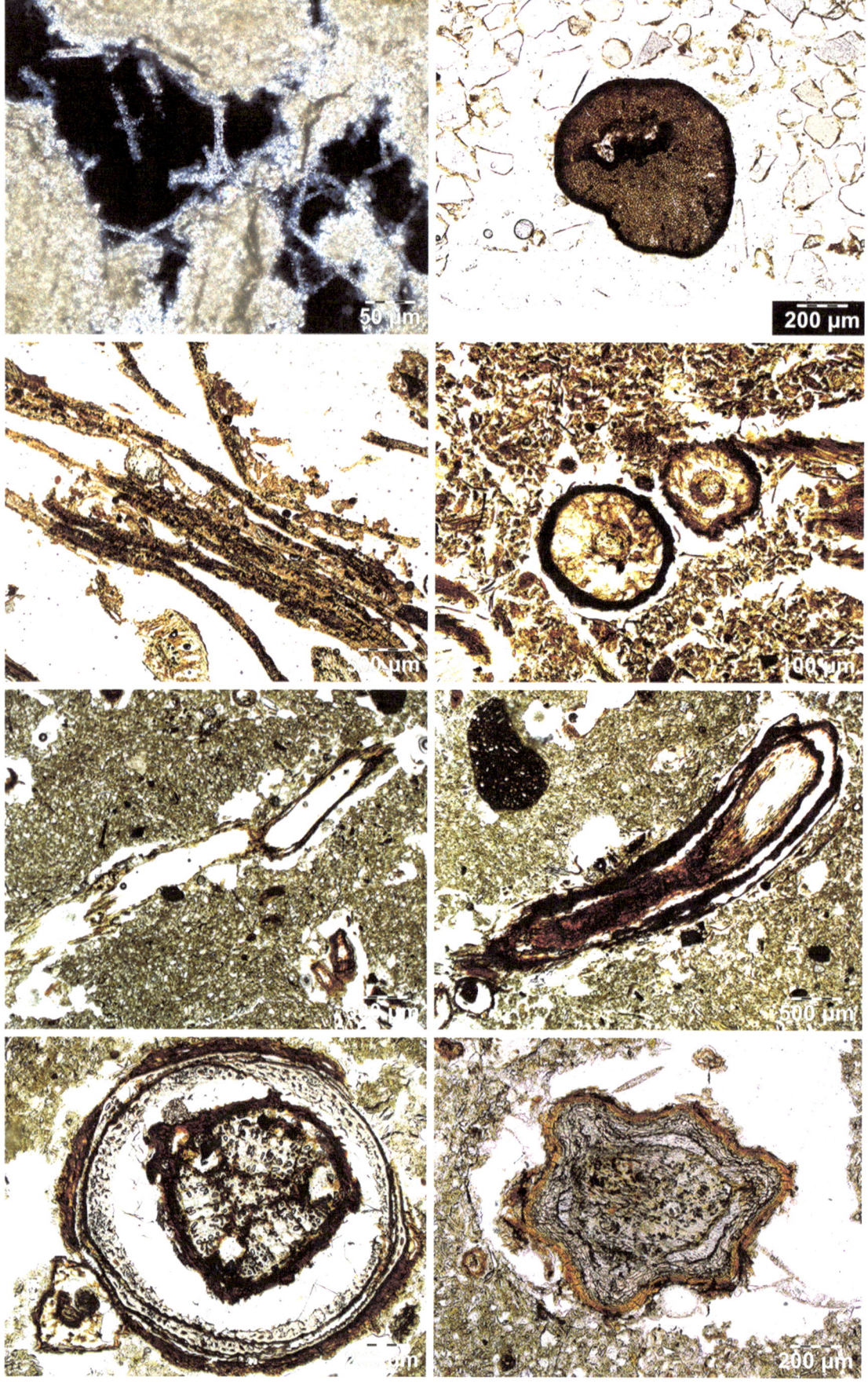

File 41: Organic Matter I

Organic matter is a common feature of soils worldwide. Soil organic matter includes remains of all plant parts, as well as small animals (insects, arthropods, etc.), fungi, and bacteria. Organic horizons are characterized by their dark colour caused by the melanization (darkening) of the soil organic matter as well as the formation of specific organic molecules. Plants are mainly composed of lignin and cellulose, giving some organic material a birefringence belonging to the first order colours in XPL. In this section, fungal features are described as well as plant material from leaves to roots.

Captions from upper left corner to lower right corner.

1. Fungal filaments covered by oxalate crystals in a Calcisol, Nazareth, Israel. The groundmass is composed of micrite. Such filaments are covered by weddellite, a hydrated calcium oxalate (at high magnification, the shape, size, and the way the crystals are attached to the filament are good clues to identify such oxalates), which is frequently confused with calcium carbonate. Therefore, the terms "calcified filaments" should be avoided, as they refer to calcium carbonate and not oxalate (Bindschedler et al. 2016). XPL view.

2. Fungal sclerotium in an A horizon from a podzol (L'Isle-Adam forest, France). Sclerotia are brown objects with sharp limits and composed of multiple and agglomerated small rounded cells. The border is generally thick and can be broken, opening the sclerotium. The groundmass is composed of sandy quartz grains arranged in a chitonic c/f related distribution. PPL view.

3. Cross-sections of leaves in a humus sample (O horizon from a Brunisol, Switzerland). These organ fragments are composed of various thin elongated and brown tissue types, i.e. the cuticle and the mesophyll. PPL view.

4. Cross-section of stems in an Ah horizon from a calcaric Leptosol (Jura Mountains, Switzerland). The dark brown border corresponds to the epidermis and the sclerenchyma of a dicot plant. The groundmass is made of large proportions of brown gel-like organic material. PPL view.

5.–6. Longitudinal sections of roots in a loamy Leptosol, northern Apennines, Italy. Only the external part (epidermis) is preserved in 5., whereas the epidermis and some cortex cells are visible in 6. Some decaying organic material is partly melanized (dark to black parts), and root disappearance after total decomposition can form channel voids. PPL view.

7.–8. Cross-sections of roots in a loamy Leptosol, northern Apennines, Italy. The internal structure of the roots is exceptionally well preserved, as it is possible to see the epidermis, the cortex cells, the pericycle, as well as the endodermis. PPL view.

File 42: Organic Matter II

Roots are the most common organic residues observed in soils. Roots not only appear differently depending on the angle at which they were cut (see "File 41"), but they can also be differentiated according to their preservation or decay. Three examples of similar root sections are given for three different ways of decaying and preservation. In addition, it is possible to find a biogenic product, starch, preserved in soils at certain locations, e.g. in arid zone soils.

Captions from upper left corner to lower right corner.

1.   Longitudinal section of a root epidermis showing the network of its cells in a brown colour due to the presence of lignin. PPL view.
2.   Same view as in 1. in XPL. The organic material has a low birefringence. The soil groundmass is scarcely enriched in calcium carbonate.
3.   Longitudinal section of a root epidermis after total darkening. The organic matter appears in dark black (melanization) but preserves the structure of the root tissue. A loess-like deposit constitutes the groundmass. Loess Plateau, Baoji, north central China.
4.   Same view as in 3. in XPL. The root mass appears completely extinct.
5.   Example of a root in which cell vacuoles have been replaced by calcite. These calcified root cells (Jaillard et al. 1991) are common features of soils and paleosols. Here, the preserved cell morphologies correspond to epidermal tissues (Jaillard et al. 1991). Calcified root cells form when the vacuole is infilled by a calcite crystal, which replaces the entire cell, fossilizing the root tissue. A loess-like deposit constitutes the groundmass. Loess Plateau, North Central China.
6.   Same as in 5. in XPL. The polarizing colours correspond to calcite.
7.–8.   Dung deposit from the central Sahara (Libya), in which organic compounds are extremely well preserved, including storage tissue cells. In these cells, starch is present as spheres that are perfectly visible in PPL and XPL. In XPL, the spheres show a black extinction cross but in an abnormal greyish birefringence colour.

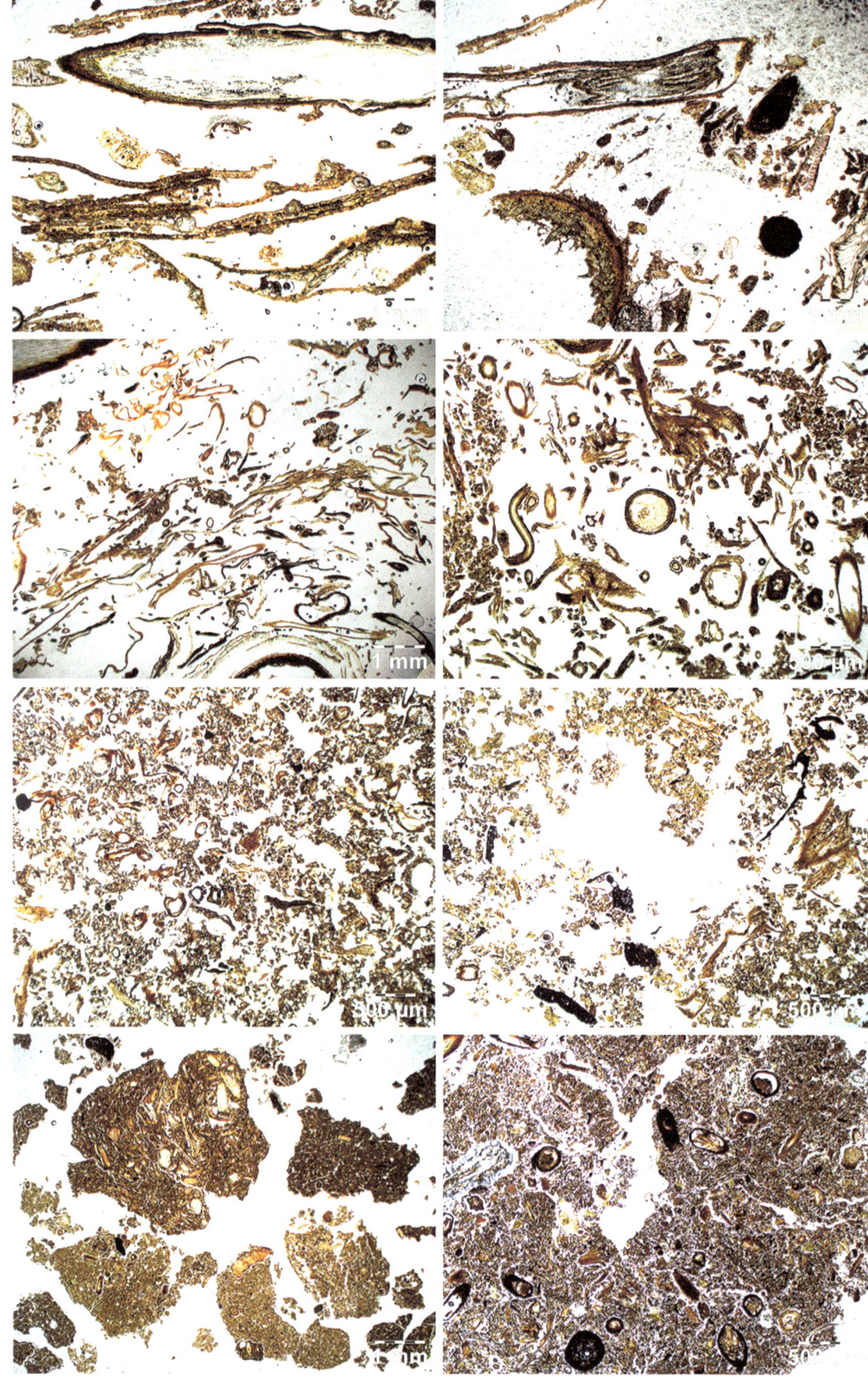

File 43: Humus

Humus forms result from decomposed organic matter lying at the surface of the soil, or present in the uppermost 30 cm. It mainly consists of combinations of Oi, Oe, Oa, and A horizons (IUSS and Working-Group-WRB 2014). The state of preservation and decomposition of plant tissues allows the type of humus form to be recognized. Humus forms are also often associated with earthworm casts and faecal pellets of small animals. A progression from litter to A horizon is presented in this section made from thin sections from a folic Umbrisol and a calcaric Leptosol, Côte de Ballens, Jura Mountains, Switzerland.

Captions from upper left corner to lower right corner. All views are in PPL.

1.–2. Thin sections of an Oi horizon showing cross-sections of leaves, twigs, and tissue fragments of a branch or bark. The organic material is not, or only very slightly, decomposed and made of large plant parts at this scale.

3. Thin section of an Oe horizon, where organic materials are fragmented by soil mesofauna. Some larger tissues or organs are still recognizable, but most of the objects are cut into small pieces and mainly represent the organic layers most resistant to decay. The comparison with 1. and 2. at the same scale emphasizes the decreasing size of the organic material.

4. Thin section of an Oe horizon with the same general characteristics as in 3. In the centre, cross-section of a stem. Some small aggregates of unrecognizable organic matter can be seen on the left and right edges of the microphotograph.

5. Thin section of an Oa horizon in which most of the organic material can no longer be recognized. Humification processes are at work. Only a few resistant or newly incorporated pieces of plant tissues still appear as continuous shapes.

6. Thin section of an Oa horizon with the same general characteristics as in 5. The only difference is the presence of small black pieces of charcoal, and close to the right edge, a triangular preserved tissue.

7. Thin section of an A horizon showing organomineral aggregates, forming crumb peds. The differences in the darkness are related to the degree of incorporation of humified material. The largest crumb includes some moderately decayed organic matter.

8. Detail of a crumb aggregate made by an earthworm in an A horizon. The groundmass is composed by a micromass of humified organic matter (as granular micro-aggregates) with the presence of some moderately decomposed plant tissues. A fungal sclerotium can be seen at the bottom and centre-left part of the picture.

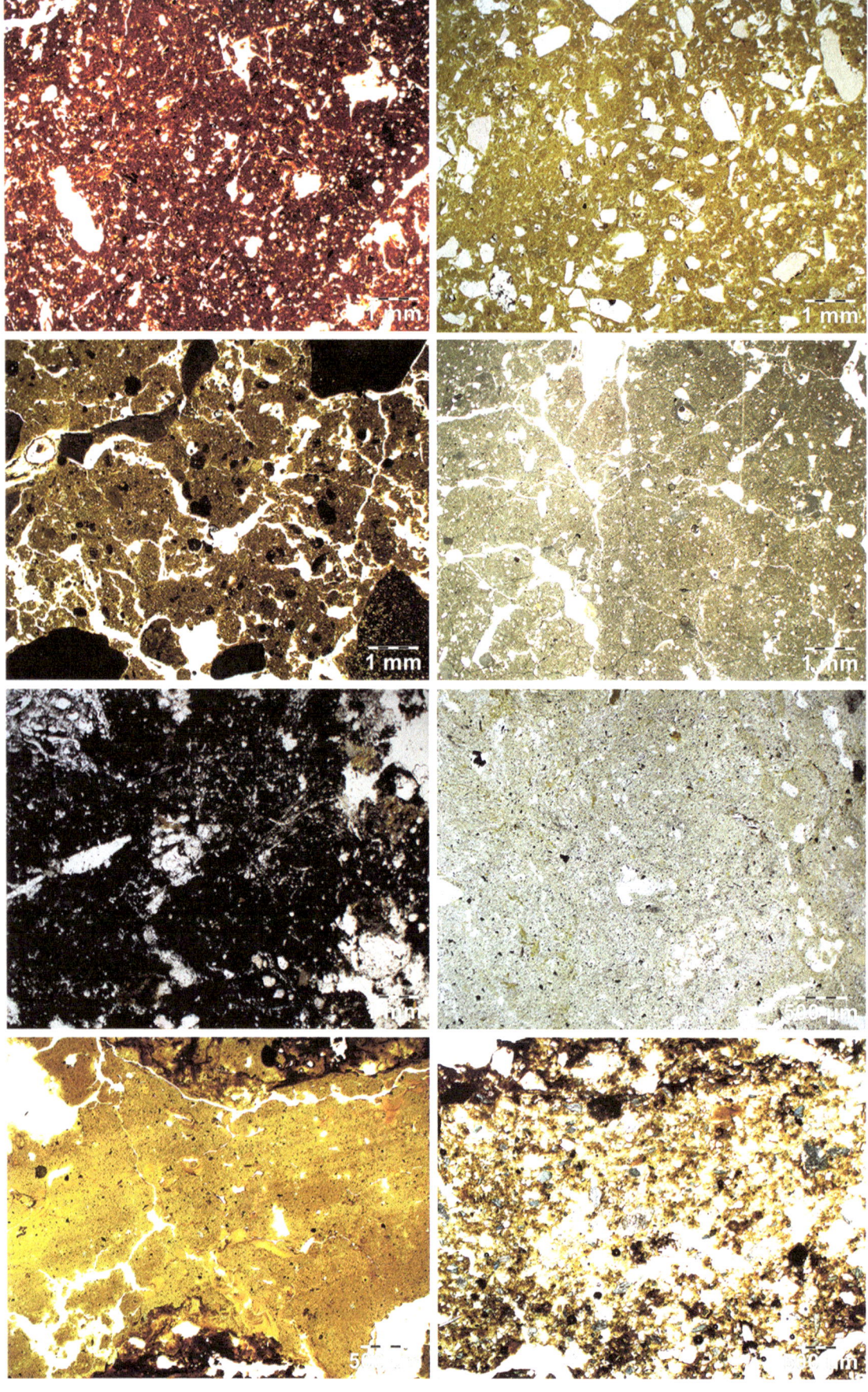

File 44: Micromass

> According to Bullock et al. (1985), micromass is a general term used to denote the finest material of the groundmass. Stoops (2003, 2021) describes the micromass as being characterized by the presence of crystalline or amorphous clay minerals, associated with oxyhydroxides or not, amorphous organic matter, and possibly the presence of small crystals of calcite (i.e. micrite) or mica. Micromass can be described using its colour, its transparency (i.e. its limpidity), and the interference colours (i.e. its b-fabric). Limpidity ranges from limpid to opaque, with intermediate states such as cloudy, speckled, and dotted. This section presents examples or micromass colours and limpidity, which have to be observed in PPL. In addition, the next two sections show examples of b-fabrics observed in XPL.

Captions from upper left corner to lower right corner.

1. Red micromass: the red colour is due to the presence of hematite mixed with clays. Chromic Luvisol, Liguria, Italy.
2. Yellow micromass: the bright yellow colour is due to the presence of goethite mixed with clay minerals. Lixisol, tropical Africa.
3. Very dark brown micromass: the brown colour is due to the presence of iron and organic matter forming the clay-humic complex. Luvisol, northern Apennines, Italy.
4. Brownish grey and cloudy micromass: the pale colour corresponds to a calcite-rich (micrite) micromass. Calcaric Leptosol, Jura Mountains, Switzerland.
5. Black and opaque micromass from a peat soil developed in central Syria. Melanized organic matter gives the black colour.
6. Greyish and limpid micromass: the reduction processes removed all oxidized iron, giving the micromass this light and limpid aspect. Gleysol formed on glacio-lacustrine clays, Jura Mountains, Switzerland.
7. Yellow and cloudy micromass: the bright yellow colour is due to the presence of goethite. Ferralsol from tropical Africa.
8. Brownish and speckled micromass: the micromass is not uniform, compared to the other examples. Cambisol, Apulia, Italy.

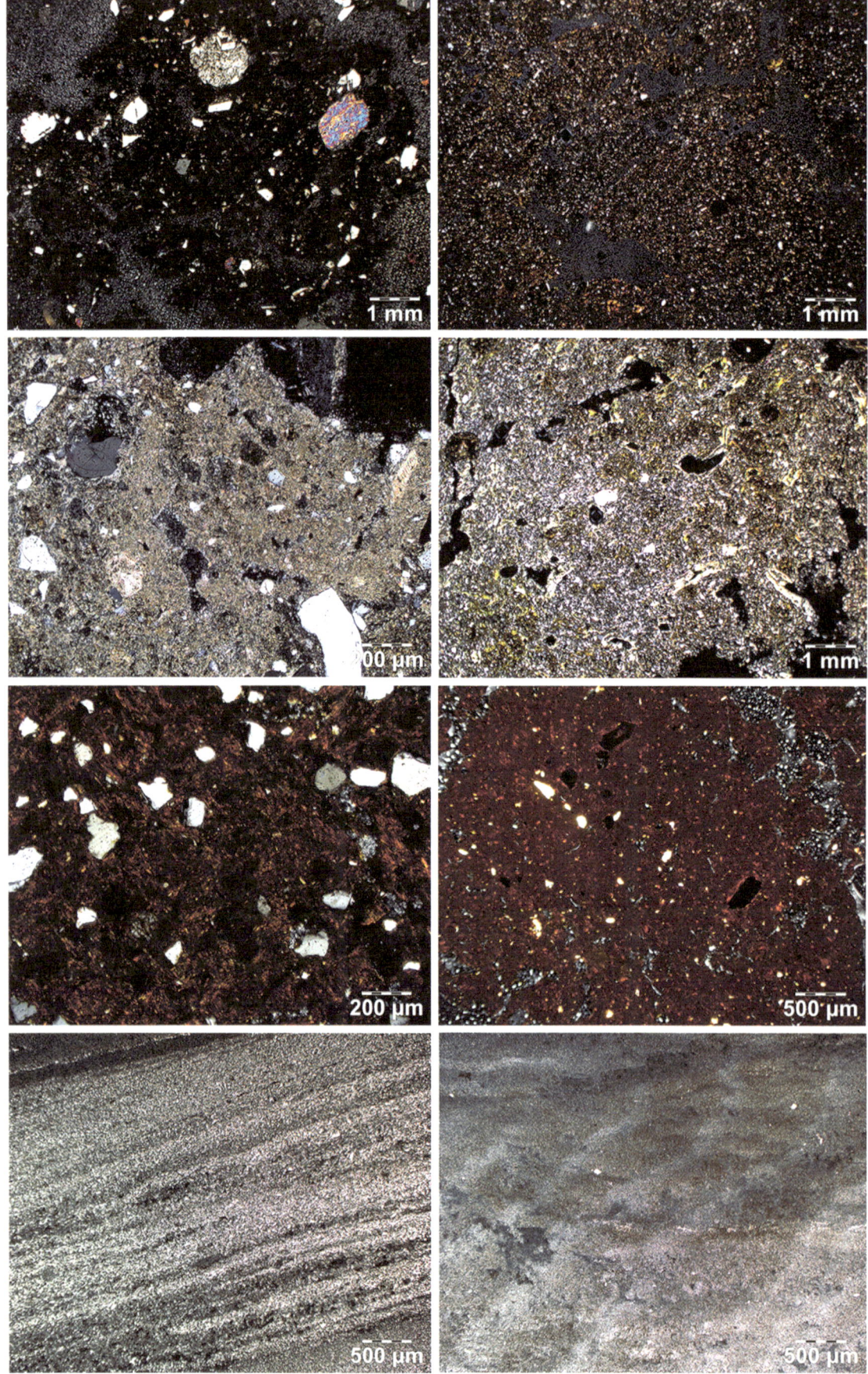

File 45: B-Fabric I

According to Bullock et al. (1985), micromass is a general term used to indicate the finest material of the groundmass. Stoops (2003, 2021) describes the micromass as being characterized by the presence of crystalline or amorphous clay minerals, associated with oxyhydroxides or not, amorphous organic matter, and possibly the presence of small crystals of calcite (i.e. micrite) or mica. Micromass can be described using its colour, its transparency (i.e. its limpidity), and the interference colours (i.e. its b-fabric). The b-fabric describes "the origin and patterns of orientation and distribution of interference colours in the micromass" (Bullock et al. 1985). This section shows examples of thin sections observed only in XPL.

Captions from upper left corner to lower right corner.

1. Undifferentiated b-fabric induced by the dominance of short-range order minerals, e.g. allophanes, in a volcanic soil (Andosol). The two largest minerals are olivine and amphibole crystals.
2. Undifferentiated b-fabric induced by a mass of fine crystals of oxyhydroxide. Paleosol (paleo-Oxisol), central Sahara, Libya.
3. Crystallitic b-fabric made of small birefringent calcite (micrite). Calcaric Leptosol, Jura Mountains, Switzerland.
4. Crystallitic b-fabric composed of large amounts of very fine mica crystals. Soil developed on loess, northern Italy.
5. Mosaic-speckled b-fabric with randomly arranged clusters of oriented clays resulting in a mosaic-like pattern. Paleosol (paleo-Luvisol), central Sahara, Libya.
6. Stipple-speckled b-fabric with randomly arranged but isolated clusters of oriented clays. Chromic Luvisol, tropical Africa.
7. Unistrial b-fabric formed by a parallel oriented micromass and displaying a typical imprint of a sedimentary pattern. Fluvisol, Rhône Valley, Switzerland.
8. Bistrial b-fabric formed by two preferred directions of orientation in the micromass. It is typical of a sedimentary pattern. Fluvisol, Rhône Valley, Switzerland.

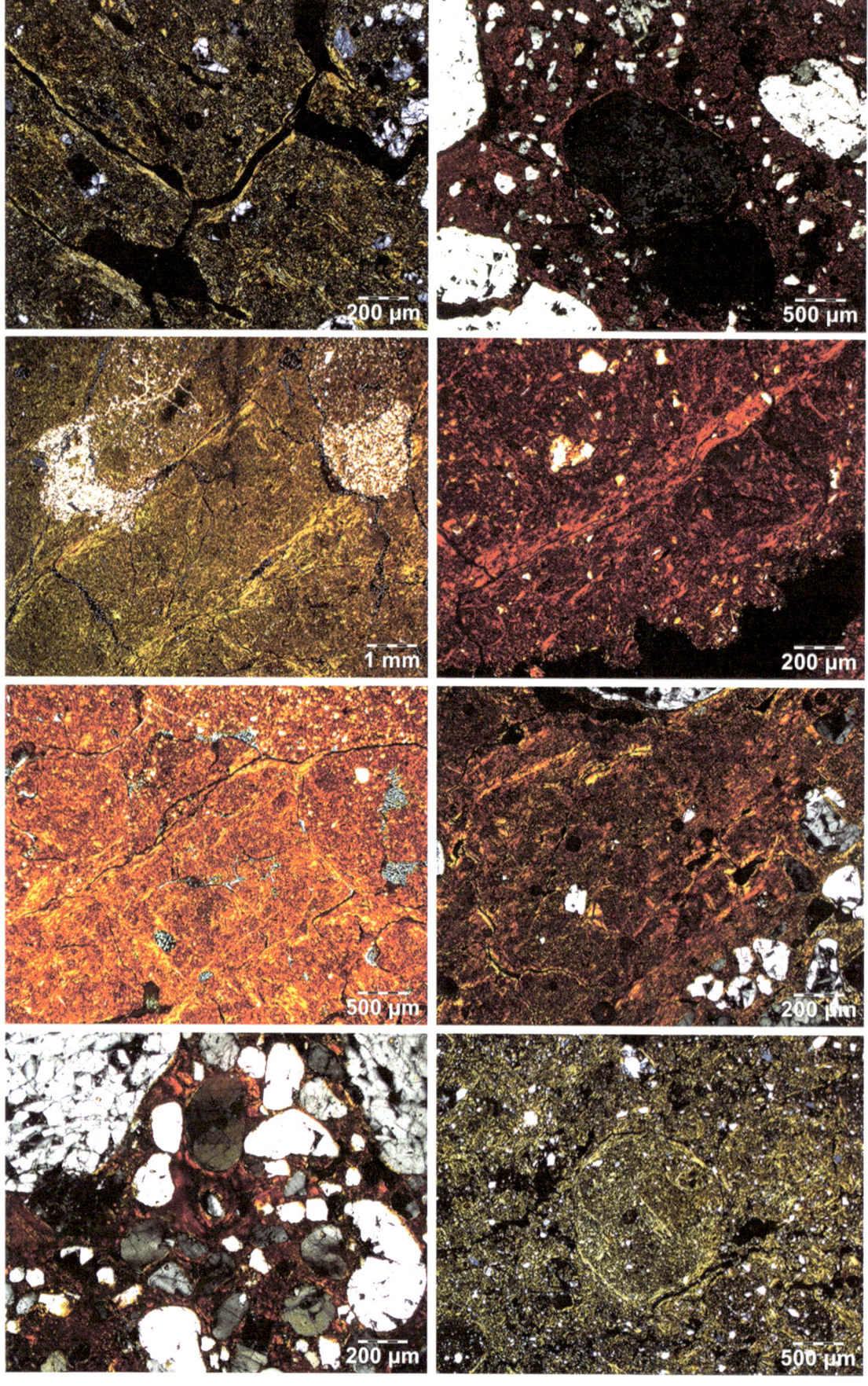

File 46: B-Fabric II

According to Bullock et al. (1985), micromass is a general term used to indicate the finest material of the groundmass. Stoops (2003, 2021) describes the micromass as being characterized by the presence of crystalline or amorphous clay minerals, associated with oxyhydroxides or not, amorphous organic matter, and possibly the presence of small crystals of calcite (i.e. micrite) or mica. Micromass can be described using its colour, its transparency (i.e. its limpidity), and the interference colours (i.e. its b-fabric). The b-fabric describes "the origin and patterns of orientation and distribution of interference colours in the micromass" (Bullock et al. 1985). This section shows examples of thin sections observed only in XPL, displaying the presence of elongated areas in which clays are approximately simultaneously extinct.

Captions from upper left corner to lower right corner.

1. Porostriated b-fabric in which a clayey assemblage is distributed parallel to the surface of pores. Chromic Paleo-Luvisol, Sardinia, Italy.
2. Granostriated b-fabric in which a clayey assemblage is distributed parallel to the surface of quartz grains. Chromic Paleo-Luvisol, Sardinia, Italy.
3. Parallel striated b-fabric showing a parallel arrangement of clayey assemblages. Chromic Paleo-Luvisol, Sardinia, Italy.
4. Monostriated b-fabric showing an isolated and independent alignment of clays in the micromass. Chromic Paleo-Luvisol, northern Italy.
5. Cross-striated b-fabric showing the intersections of multiple sets of clay alignments. Chromic Luvisol, Liguria, Italy.
6. Random striated b-fabric does not show any organization in the arrangement of clay alignments, giving an irregular pattern. Chromic Paleo-Luvisol, Sardinia, Italy.
7. Concentric striated b-fabric displaying multiple concentric rings of oriented clay material. Paleosol, Libya, central Sahara.
8. Circular striated b-fabric in which a clayey assemblage is distributed in a circular arrangement, for instance in a ring shape. Luvisol, Apulia, Italy.

Chapter 4

Pedogenic Features

File 47: Imprints of Pedogenesis

From a historical point of view, soil micromorphology was first used in order to decipher the expressions of pedogenic processes at the microscale (Kubiëna 1938). In the preceding chapters, the Atlas listed a series of descriptive tools to help with the identification of objects. This chapter deals with specific pedofeatures encountered in a large diversity of soils and directly related to pedogenic processes. Pedological features (Brewer 1964) or pedofeatures (Bullock et al. 1985) are "discrete fabric units present in soil materials that are recognizable from an adjacent material by a difference in concentration in one or more components or by a difference in internal fabric" (Stoops 2003, 2021). In Stoops (2003, 2021), pedofeatures are subdivided into two categories: matrix pedofeatures and intrusive pedofeatures. Matrix pedofeatures can be subdivided according to their relationship with the groundmass (depletion, impregnative, and fabric pedofeatures) and to their morphology (hypocoatings, quasicoatings, matrix infilling, intercalation, and matrix nodules). Regarding the intrusive pedofeatures, they include coatings, infillings, crystals and crystal intergrowth, intercalations, and finally nodules. The proposed nomenclature of this chapter is based on the nature and morphology of the pedofeatures, simplified from Bullock et al. (1985).

Examples of pedofeatures from left to right: a polygenetic nodule (XPL), hypo- and quasicoatings (PPL), clay infilling (PPL), infilling of needle-fibre calcite (XPL), and pellets (PPL).

E. P. Verrecchia, L. Trombino, *A Visual Atlas for Soil Micromorphologists*,
https://doi.org/10.1007/978-3-030-67806-7_4

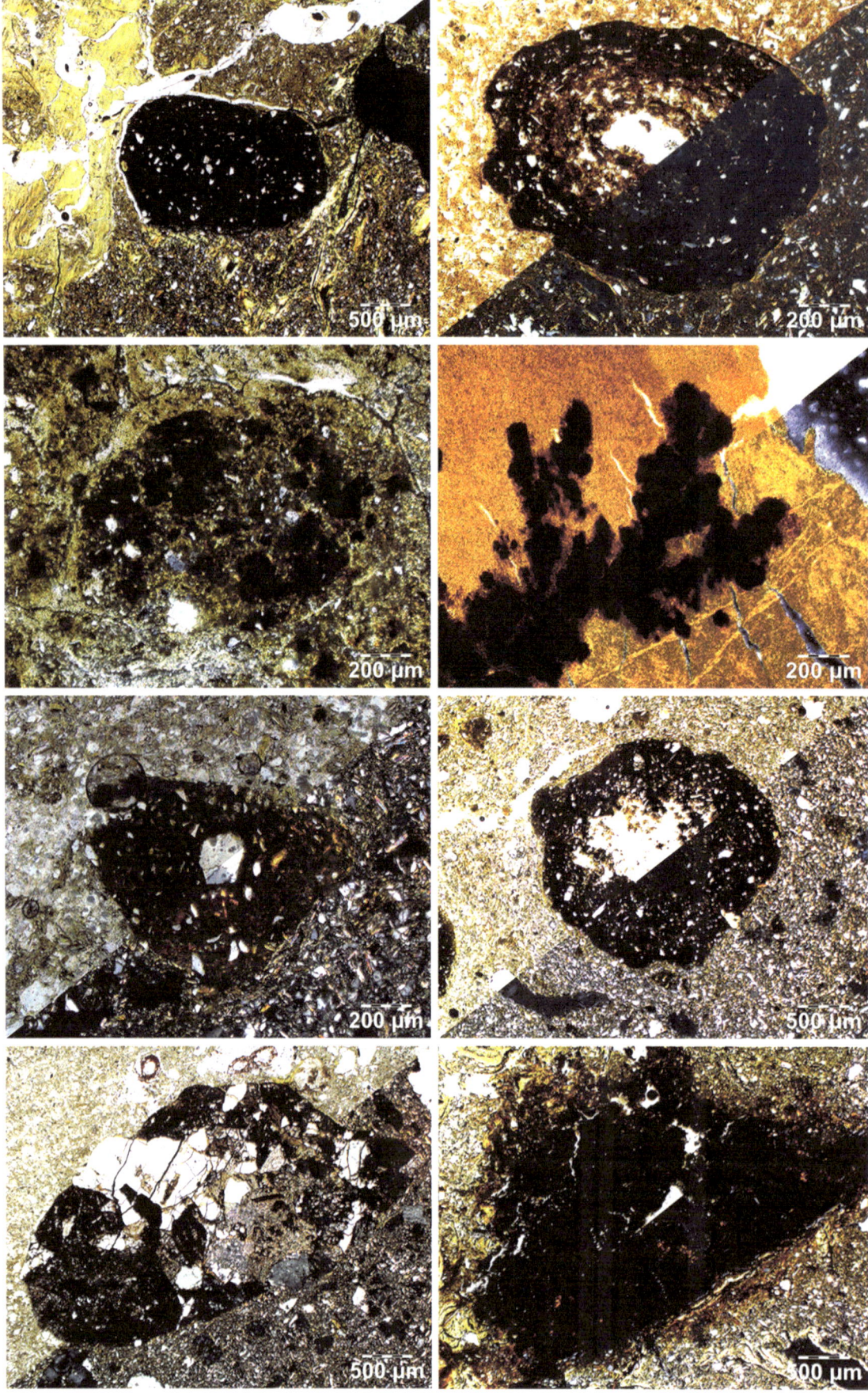

File 48: Iron- and Manganese-Bearing Nodules

Nodules are defined as roughly equidimensional pedofeatures that are not related to natural surfaces or voids and that do not consist of single crystals. From a theoretical point of view, nodules can be regarded as matrix impregnative or intrusive pedofeatures (Stoops 2003). This plate presents iron-bearing nodules, and according to Bullock et al. (1985), they can be classified as amorphous or cryptocrystalline pedofeatures based on their internal fabric and external morphology. The chemical nature of nodules is often confirmed using an electron microprobe; for example, element mapping frequently shows the association of iron and manganese (see "File 7").

Captions from upper left corner to lower right corner.

1. Typic nodule: iron-bearing nodule with undifferentiated internal fabric and sharp boundaries in a loamy groundmass. Paleo-Luvisol, Piedmont region, Italy.
2. Concentric nodule: iron-bearing nodule with concentric layers of matter and sharp boundaries in a sandy–clayey groundmass. Chromic Luvisol, Apulia, Italy.
3. Aggregate nodule: such nodules are formed by an aggregation of small mostly typic nodules and are frequently found in Vertisols. Vertisol, Po Plain, Italy.
4. Dendritic nodule: Stoops (2003, 2021) considers them to be aggregate Fe–Mn nodules organized in a dendritic pattern. This dendritic nodule developed within a clayey micromass. Chromic Paleo-Luvisol, Sardinia, Italy.
5. Nucleic nodule: example of a Fe–Mn nodule precipitated around a quartz grain, i.e. an allochthonous core, in a loamy groundmass. Oxyaquic Cryosol, Italian Alps.
6. Geodic nodule: nodule showing an empty core, i.e. an irregular-shaped void in a silty groundmass. Loess paleosol, central Po Plain, Italy.
7.–8. Alteromorphic nodules: this type of nodule is usually the product of weathering. These nodules are also characterized by pseudomorphosis of mineral or organic materials. In 7., alteromorphic nodule developed from a rock fragment in a loess paleosol. Ligurian Alps, Italy. In 8., alteromorphic nodule incorporating plant residue. Paleo-Luvisol, Piedmont region, Italy.

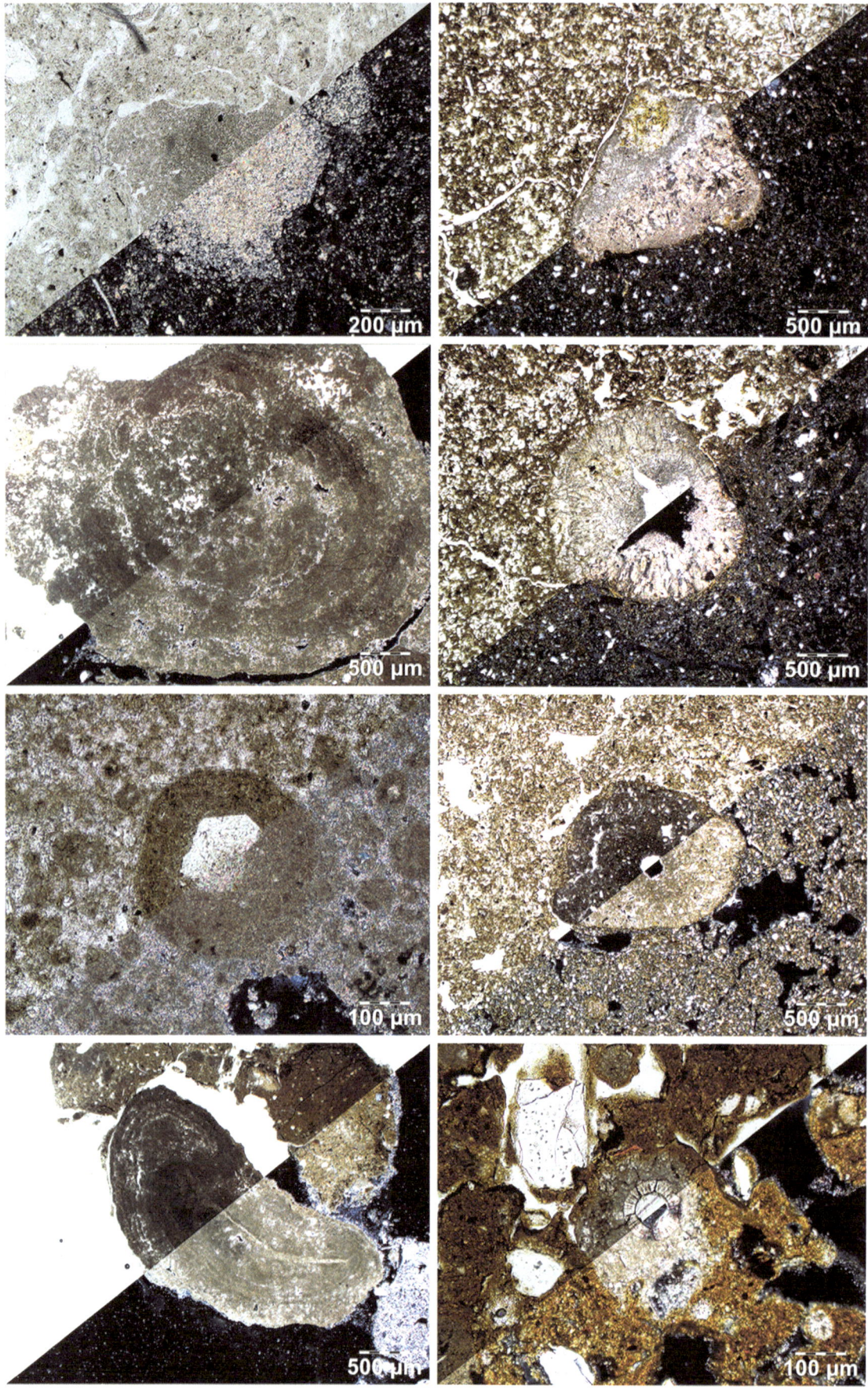

File 49: Carbonate Nodules

Nodules are defined as roughly equidimensional pedofeatures that are not related to natural surfaces or voids and do not consist of single crystals (Stoops 2003). From a theoretical point of view, nodules can be regarded as matrix impregnative or intrusive pedofeatures (Stoops 2003). This plate presents carbonate nodules, and according to Bullock et al. (1985), they can be classified as crystalline pedofeatures based on their internal fabric and external morphology. Moreover, the size of carbonate crystals forming the nodule (i.e. micrite, microsparite, or sparite) is also a pertinent attribute of such pedofeatures.

Captions from upper left corner to lower right corner.

1. Typic nodule: micritic nodule with a homogeneous internal fabric and sharp boundaries in a carbonate micromass. Cambisol, Paris Basin, France.
2. Sparitic nodule: carbonate nodule composed of large sparitic crystals and sharp boundaries (see "File 51") in a silty–clayey groundmass. Planosol, Po Plain, Italy.
3. Concentric nodule: such nodules are formed by an accretion of concentric layers of micrite. This particular nodule can result from precipitation in a swampy environment. Calcisol (Gleyic), Syria.
4. Septaric nodule: the void in the centre results from radiating cracks. This septaric nodule developed within a silty–clayey groundmass. They are also frequently observed in tropical Vertisols. Planosol, Po Plain, Italy.
5. Nucleic nodule: example of a micritic nodule precipitated around a quartz grain, i.e. an allochthonous core, in microspartic and micritic groundmass. Calcisol, Negev Desert, Israel.
6. Geodic nodule: nodule showing an empty core, i.e. a round void in this case, in a silty groundmass. Cambic Calcisol, Jura Mountains, Switzerland.
7.–8. Lithoclasts and not nodules: the shape of some carbonate lithoclasts can be confused with pedogenic nodules. In 7., example of a travertine oncoid from a Calcisol (Gleyic), Syria. In 8., lithoclast of a marine limestone with a foraminifera fragment (sparitic round feature in the centre) in a Chromic Cambisol, Apulia, Italy.

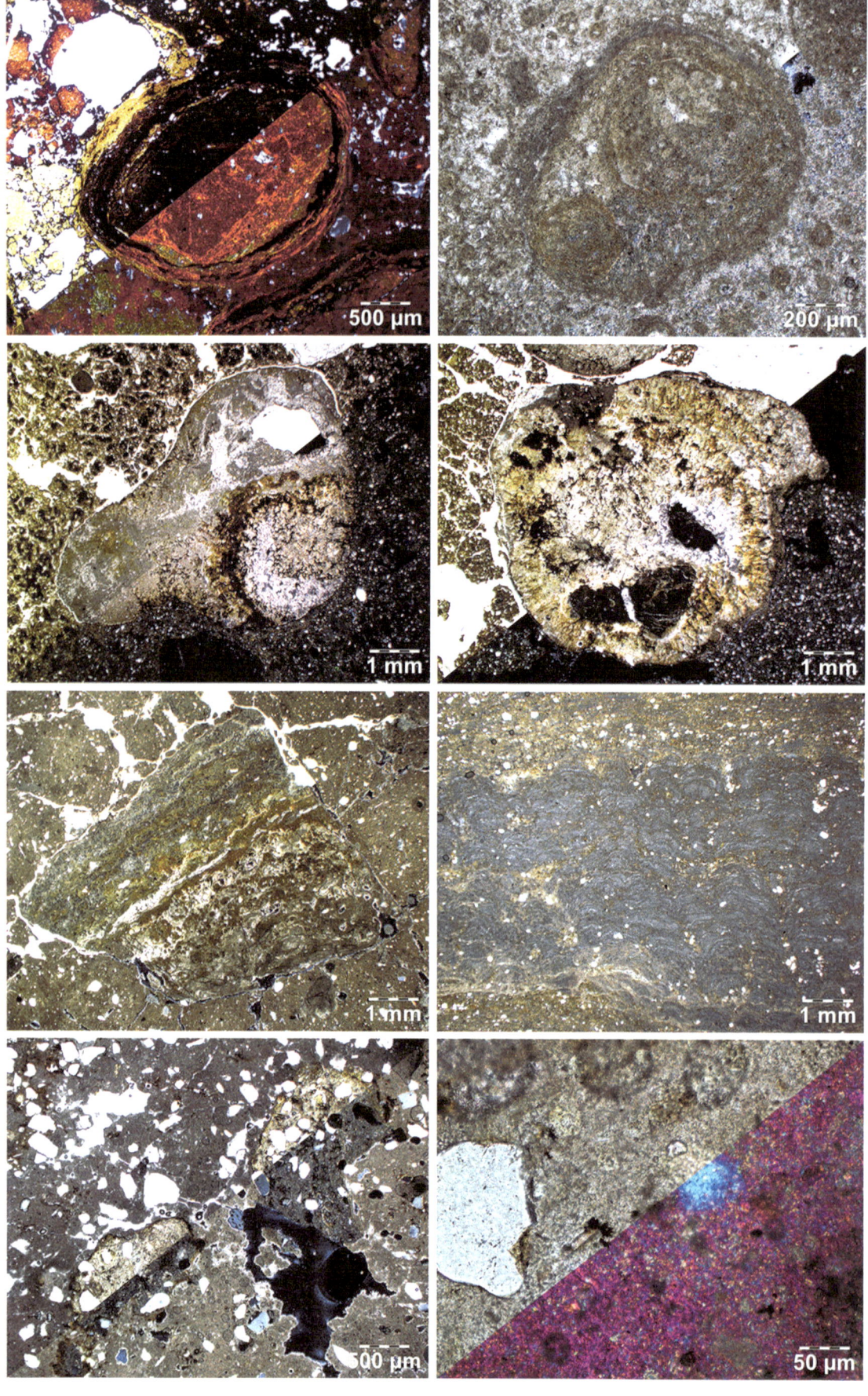

File 50: Polygenetic Nodules

Polygenetic nodules are either nodules composed of multiple generations of cortical layers or products of different pedogenetic phases. A special type of polygenetic nodule is associated with perlitic crusts. These crusts are made of multiple polygenetic nodules, also called ooids (Durand et al. 2018), of various sizes in a monic to close porphyric c/f related distribution and almost always associated with carbonate laminar crusts.

Captions from upper left corner to lower right corner.

1.   Polygenetic multi-layered iron-bearing nodule. The internal part is composed of a large nucleus made of iron oxyhydroxides impregnating the groundmass and clays. This nucleus is detached from a cortex formed by multiple thin layers of the same mineralogical composition. Paleosol, Liguria, Italy.

2.   Polygenetic carbonate nodule composed of smaller nodules surrounded by a polygenetic micritic cortex in a micromass from a perlitic crust. Calcisol, Negev Desert, Israel.

3.   Polygenetic carbonate nodule formed, first, by a sparitic typic nodule, rimmed by an iron oxyhydroxide brown layer and aggregated into a larger partially geodic and micritic nodule. Vertisol, Extreme North region, Cameroon.

4.   Polygenetic sparitic nodule. The void in the centre is surrounded by clear sparitic cement, whereas the central part of the nodule consists of sparite with traces of iron oxyhydroxides. The cortex is also calcitic with traces of iron oxyhydroxides, but crystals are organized in a fan-like arrangement. Such nodules are frequently due to water-table fluctuations. Vertisol, Extreme North region, Cameroon.

5.–6.   In 5., fragment of a reworked laminar crust included in a Calcisol (Gleyic), Syria. This feature is not strictly considered as a "pedogenic" nodule, but the fact that it originates from the same environment as the groundmass makes it polygenetic nevertheless, because it has been eroded, displaced, and deposited a bit further from its place of formation. In 6., *in situ* laminar crust showing the alternation of clear greyish microsparitic and yellowish–brownish micritic layers. The microsparitic layers are made of calcitic spherulites (see "File 37" and Verrecchia et al. 1995). PPL view, Negev Desert, Israel.

7.–8.   Two polygenetic siliceous nodules cross-cut by the picture's diagonal. They result from the secondary silicification of primary calcitic nodules trapping quartz grains. They are also rimmed by a very thin layer of oxyhydroxides. The groundmass is composed of quartz grains in a micritic micromass. In 8., close-up. The XPL view with a gypsum plate emphasizes the abundance of silica in the micromass (high-order colours). A quartz grain appears in bright blue. Calcisol, Chobe Enclave, Botswana.

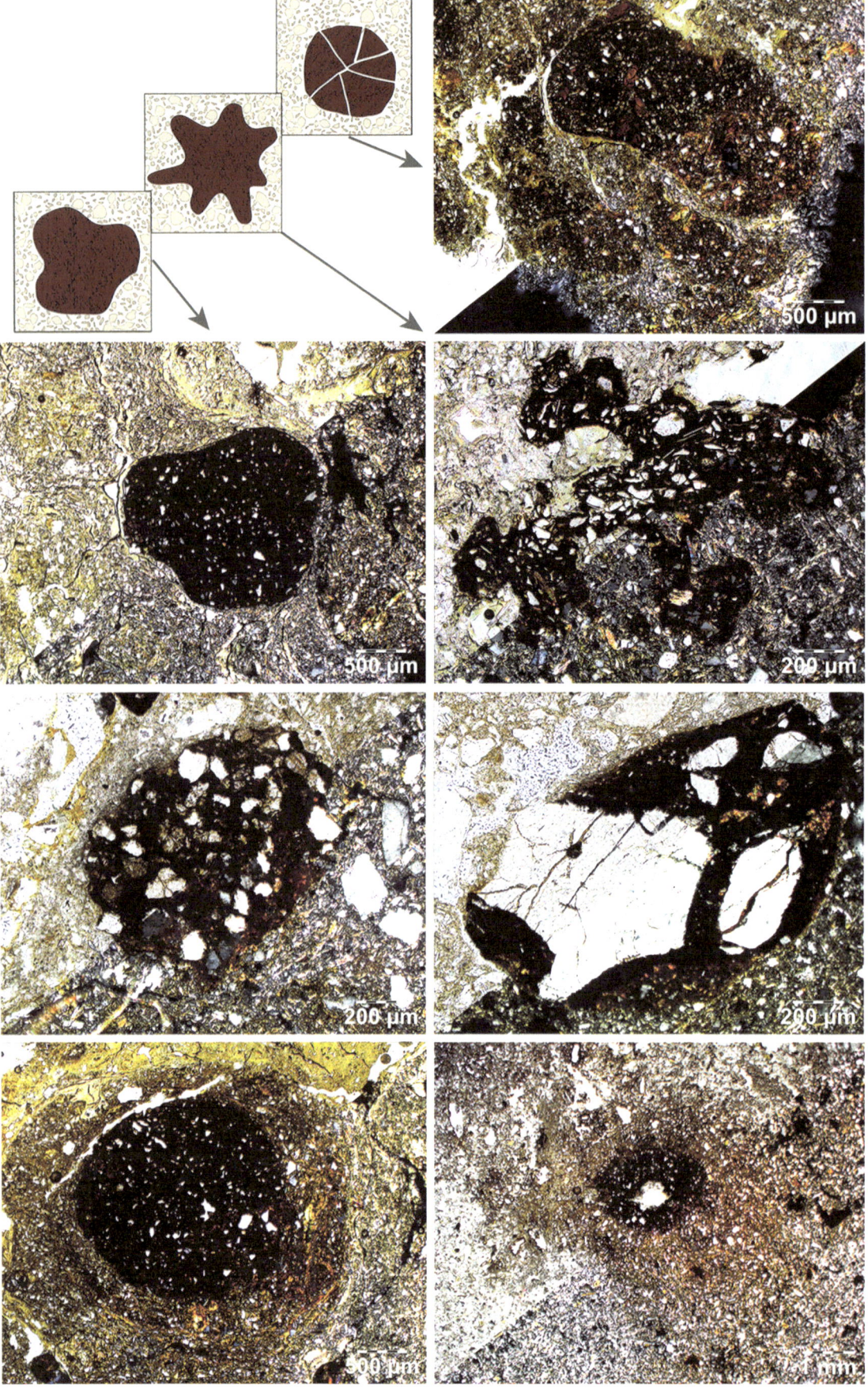

File 51: Nodules: Morphology and Border Shape

Nodules can be characterized by their specific external morphologies. As pedofeatures, nodules also have spatial relationships with the groundmass surrounding them; this relationship is illustrated by the shape of nodule's border.

Captions from upper left corner to lower right corner.

1. Sketch of the different external morphologies of nodules (Stoops 2003). The mammillate type, in the lower left corner, refers to an undulating external shape. The digitate type (centre) makes fingerlike penetrations inside the adjacent groundmass. The disjointed type, in the upper right corner, is composed of angular accommodating fragments.
2. Example of disjointed iron-bearing nodule. Paleo-Luvisol, Piedmont region, Italy.
3. Example of mammillate iron-bearing nodule. Paleo-Luvisol, Piedmont region, Italy.
4. Example of a digitate iron-bearing nodule. Paleo-Luvisol, Piedmont region, Italy.
5.–6. Example of two different iron-bearing nodules with sharp boundaries. This kind of border is often indicative of an allochthonous provenance of the nodules. Paleo-Luvisol, Liguria, Italy.
7.–8. Example of two different iron-bearing nodules with diffuse boundaries. In both microphotographs, the outer layers of the nodule diffuse into the groundmass, in an isotropic manner in 7. and as a gradually fading halo in 8. These kinds of borders are often related to an *in situ* formation process. These nodules have been observed in 7. in a Paleo-Luvisol, Piedmont region, Italy and in 8., in a Fluvisol, Jura Mountains, Switzerland.

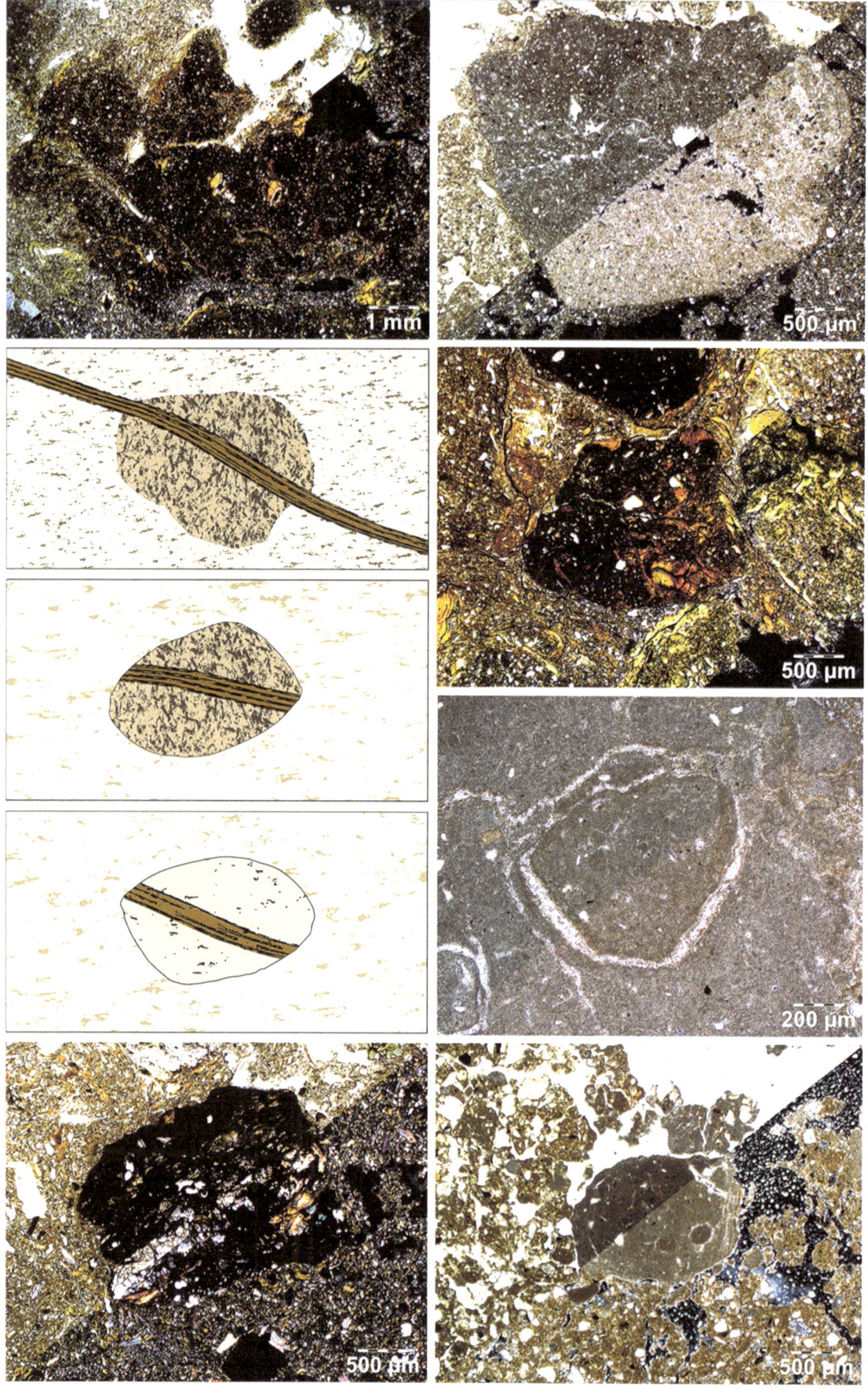

File 52: Nodules: Orthic, Anorthic, and Disorthic

Nodules can be formed in situ, reworked to varying degrees, or inherited from the parent material. In order to identify their origin, Stoops (2003, 2021) suggests the classification proposed by Wieder and Yaalon (1974). When nodules are inherited from the parent material or are clearly allochthonous, they are called *anorthic*. If nodules are formed *in situ* and do not show any sign of reworking, they are considered as *orthic*. If they were locally displaced inside the soil, they are qualified as *disorthic*.

Captions start with the sketch in the central left column. Then, the microphotographs are described clockwise around the sketch, starting from the upper left corner to the lower left corner.

1. Sketch of the three types of relationships between nodules and the surrounding soil groundmass (Stoops 2003). Top: an orthic nodule is formed *in situ*, and not displaced, as evidenced by the linear pedofeature cross-cutting the nodule, as well as the groundmass. Middle: a disorthic nodule formed inside the soil but that has been locally reworked; the linear pedofeature is only observed inside the nodule, but not in the groundmass. Bottom: the nature of an anorthic nodule is different from the groundmass and is clearly allochthonous.

2. Example of an orthic iron-bearing nodule: the coarse fraction of the groundmass (silty quartz grains) shows the same distribution pattern inside and outside the nodule, emphasizing the *in situ* concentration of soil iron compounds. Paleo-Luvisol, Piedmont region, Italy.

3. Example of an orthic nodule in a carbonate-rich environment: this orthic nodule is formed by precipitation of pedogenic micrite and microsparite. Calcisol, Jura Mountains, Switzerland.

4. Although embedding the same compounds and components as the groundmass, this disorthic nodule has been locally displaced in the profile as shown by the light orange clay coatings, which are not continuous with the groundmass. Paleo-Luvisol, Piedmont region, Italy.

5. The identification of this pedofeature as a disorthic nodule is based not only on the similar nature of the groundmass but also mostly on the presence of a brownish rim around it. It must be noted that distinguishing orthic from disorthic (but also anorthic) matrix nodules is rather difficult and sometimes impossible. Calcic paleosol, Aquitaine Basin, France.

6. Typical anorthic nodule composed of a greyish micrite and clearly different from the surrounding groundmass made of silty quartz, in a carbonate–silicate micromass. Calcisol, Syria.

7. Typical anorthic iron-bearing nodule inherited from the soil parent material and included in a fine-grained horizon. Paleosol, Liguria, Italy.

File 53: Crystals and Crystal Intergrowths

Crystals and crystal intergrowths are individual or clusters of crystals, which are precipitated inside the soil groundmass. Such crystals are not inherited from the parent material but are the product of pedogenic processes. According to Stoops (2003, 2021), crystal intergrowths are subdivided based on their distribution and/or orientation pattern, and according to Bullock et al. (1985), they can be classified as crystalline pedofeatures based on their internal fabric and external morphology.

Captions from upper left corner to lower right corner.

1. Random crystals of gypsum in a Gypsic Calcisol. Such crystal intergrowth is distributed in a random pattern and is characterized by a large variety of grain sizes. Negev Desert, Israel.
2. Parallel gypsum crystal intergrowth in a Gypsic Calcisol. XPL view, Negev Desert, Israel.
3. Goethite fan-like crystal intergrowth (orange cluster in the centre of the microphotograph). PPL view, Paleosol, central Po Plain, Italy.
4. Crystal intergrowth of vivianite organized in a radial pattern around a central point. Note the diagnostic blue colour of the vivianite crystals in PPL. Medieval archaeological soil, Po Plain, Italy.
5. Crystals of calcite developed inside a micritic groundmass, where they form sparitic intergrowths. Note the euhedral tip of some crystals. Petric Calcisol, Madrid Basin, Spain.
6.–8. Random gypsum crystal intergrowths. The variable pleochroism of the crystals shown in 6. in PPL is due to a change in mineralogy. In 7. (in XPL), gypsum crystals in various shades of grey include some other minerals with high-order interference colours, which are calcite crystals. The use of a gypsum plate in 8. clearly reveals an ongoing process of pseudomorphosis of gypsum by calcite. Gypsic Calcisol, Libya.

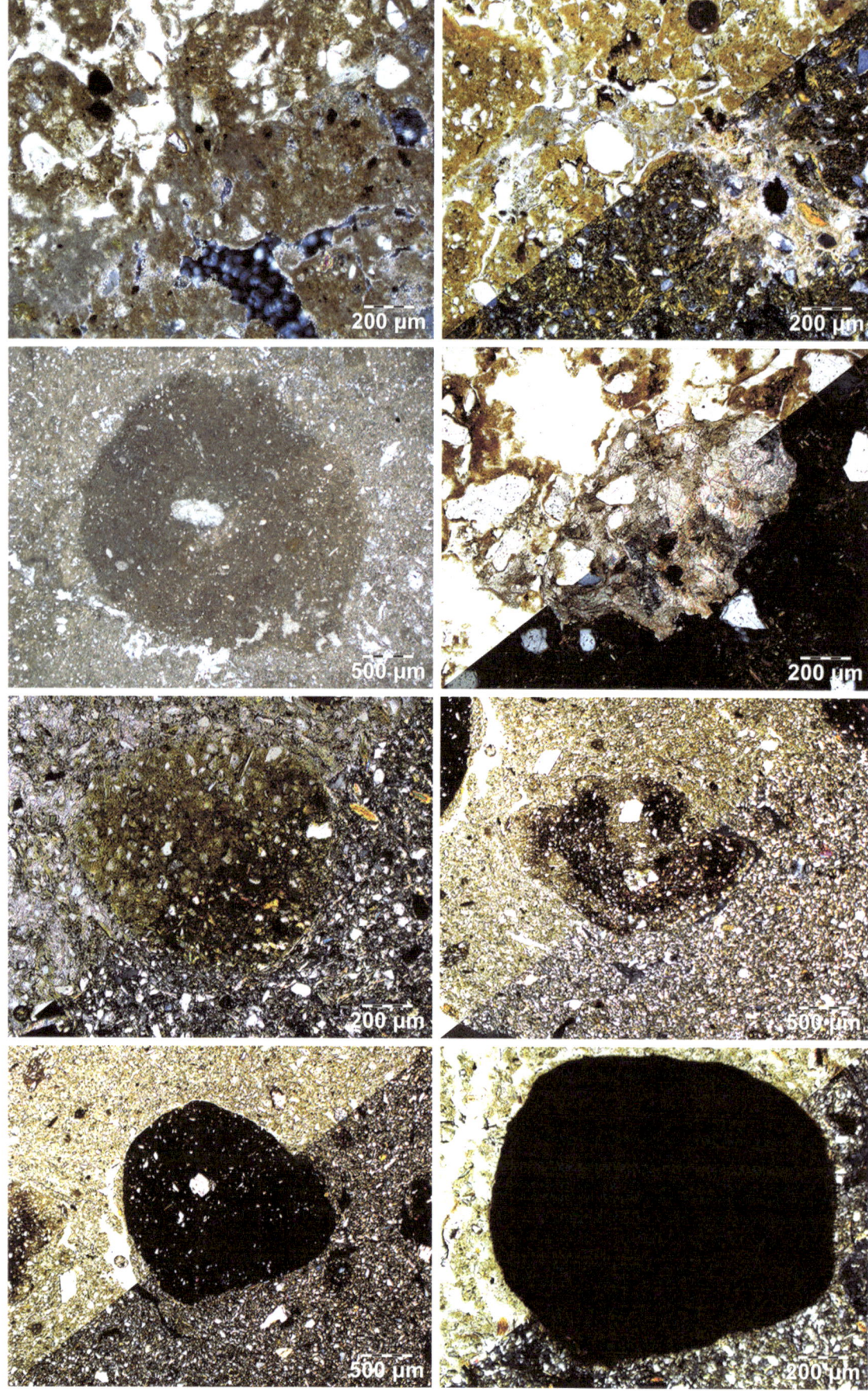

File 54: Impregnations

One of the descriptive parameters of matrix pedofeatures includes the degree of impregnation. Nodules are sometimes regarded as matrix pedofeatures (see "File 48" and "File 52"). Four degrees of impregnation are proposed by Stoops (2003, 2021) according to the amount of recognizable components of the groundmass, i.e. regarding the purity of the feature: weakly impregnated, moderately impregnated, strongly impregnated, and pure. The first four microphotographs refer to a carbonate environment, whereas the last four to aluminosilicate- and iron-rich settings.

Captions from upper left corner to lower right corner.

1. Weakly impregnated groundmass by greyish pedogenic calcite, as a first stage of carbonate nodule formation. Calcisol (Gleyic), Syria.
2. Moderately impregnated groundmass by pedogenic calcite (greyish in PPL and brownish yellow in XPL) of the size of micrite and microsparite. Components of the groundmass are still visible. Chromic Cambisol, Apulia, Italy.
3. Strongly impregnated groundmass forming a geodic nodule. The concentration of micritic matter increases significantly in the pedofeature, but the nature of the groundmass can still be identified. Fluvic Stagnosol, Jura Mountains, Switzerland.
4. Pure sparitic nodule in which the micromass of the groundmass is no longer identifiable, whereas some coarse quartz grains remain visible. Chromic Cambisol, Madrid Basin, Spain.
5.–8. Examples of progressive impregnation of the groundmass by iron oxyhydroxides forming weakly impregnated to pure typic iron-bearing nodules. Note that the components of the groundmass are progressively less recognizable, until they totally disappear. Paleo-Luvisol, Piedmont region, Italy.

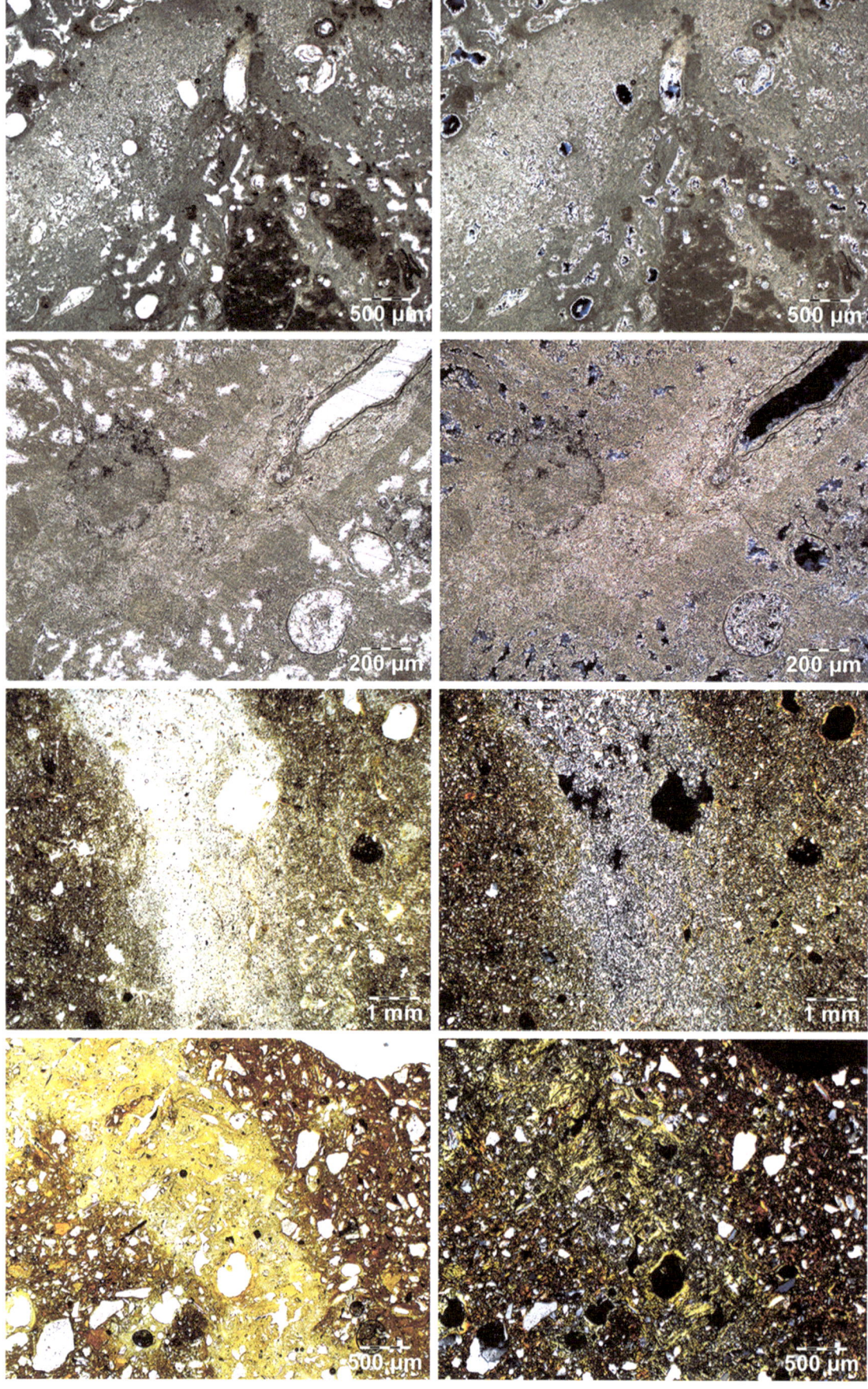

File 55: Depletions

> Depletion pedofeatures are defined as lower concentrations of a given component of the micromass, e.g. calcite or iron oxyhydroxides (Bullock et al. 1985; Stoops 2003, 2021). The loss of matter can be related to either dissolution, e.g. in a carbonate environment, or redox processes, e.g. in iron-rich environments. The mobilized ions are then translocated or leached inside or outside the profile, respectively.

Captions from upper left corner to lower right corner.

1.–2.   1. PPL view of an area depleted in carbonate emphasized by contrast and intensity of the grey colour of the micritic micromass. 2. Same view in XPL. The pore network is due to rootlets. Petric Calcisol, Galilee, Israel.

3.–4.   3. PPL view of an area depleted in carbonate induced by a rootlet. 4. Same view in XPL. The thinner calcitic crystals display interference colours of higher orders. Petric Calcisol, Galilee, Israel.

5.–6.   5. PPL view of the depleted area emphasized by contrast and intensity of the dark/light colours of the micromass. 6. Same view in XPL. The contrasted interference colours remain in XPL. Moreover, this colour variation is due to a loss of iron oxyhydroxides. Paleo-Luvisol, Piedmont region, Italy.

7.–8.   7. PPL view of an area depleted in yellow compared to the original micromass, which is reddish brown. 8. Same view in XPL. The yellow clays in the depleted area have lost the iron oxyhydroxides adsorbed on their sheets, giving them a lighter colour compared to the clays inside the original groundmass. Paleosol, central Po Plain, Italy.

File 56: Coatings with Clays I

> Coatings are defined as intrusive pedofeatures that coat natural surfaces of voids, grains, or aggregates inside soils (Stoops 2003, 2021). Coatings must not be confused with infillings (see "File 61"). Coatings are constituted by various material types, i.e. clays, and coarse, amorphous, or crystalline material. This section shows coatings formed by clays, one of the earliest pedogenic features recognized in thin sections. *In situ* clay coatings are diagnostic features of leaching processes and are also used in soil classifications. Clay coatings are described according to their colour, the presence or absence of laminations, their thickness, and their grain size (i.e. they can be called textural pedofeatures).

Captions from upper left corner to lower right corner.

1. Clay coating deposited inside a void. This thick yellow clay coating is formed by two to three layers of fine clay crystals, partly orientated in the same direction. This preferential orientation of crystals is emphasized by the observation of a large extinction band in XPL. The yellow colour can be related to the presence of goethite. Paleo-Luvisol, Piedmont, Italy.
2. Red clay coating on both sides of a channel. There is no obvious preferential orientation of the clay crystals. In addition, the red colour can be related to the presence of hematite. Paleosol, Ligurian coast, northern Italy.
3. Non-laminated yellow clay coating. Paleo-Luvisol, Piedmont, Italy.
4. Strongly laminated orange clay coating. The large succession of layers is due to the occurrence of multiple phases of clay translocation. Erosion phases can also interrupt the regularity of the layering. Paleosol, central Po Plain, Italy.
5. Limpid coating formed by very fine clay crystals. This laminated coating shows some erosional internal surfaces emphasized by concave contacts visible in XPL. Paleo-Luvisol, Piedmont, Italy.
6. Coarse-grained coating in a void. The coating is formed by the succession of layers of coarser to finer grains upwards, from silt to clay. The uppermost layer contains the largest clay fraction. Paleosol, Libya, central Sahara.
7. A typic clay coating around a transversal section of a channel showing a uniform thickness in all the directions. Chromic Luvisol, Apulia, Italy.
8. Crescent coating characterized by a larger basal thickness compared to the sides at the bottom of a void. This laminated coating displays a sharp extinction band, emphasizing the continuous parallel orientation of deposited clays. Paleosol, Ligurian coast, northern Italy.

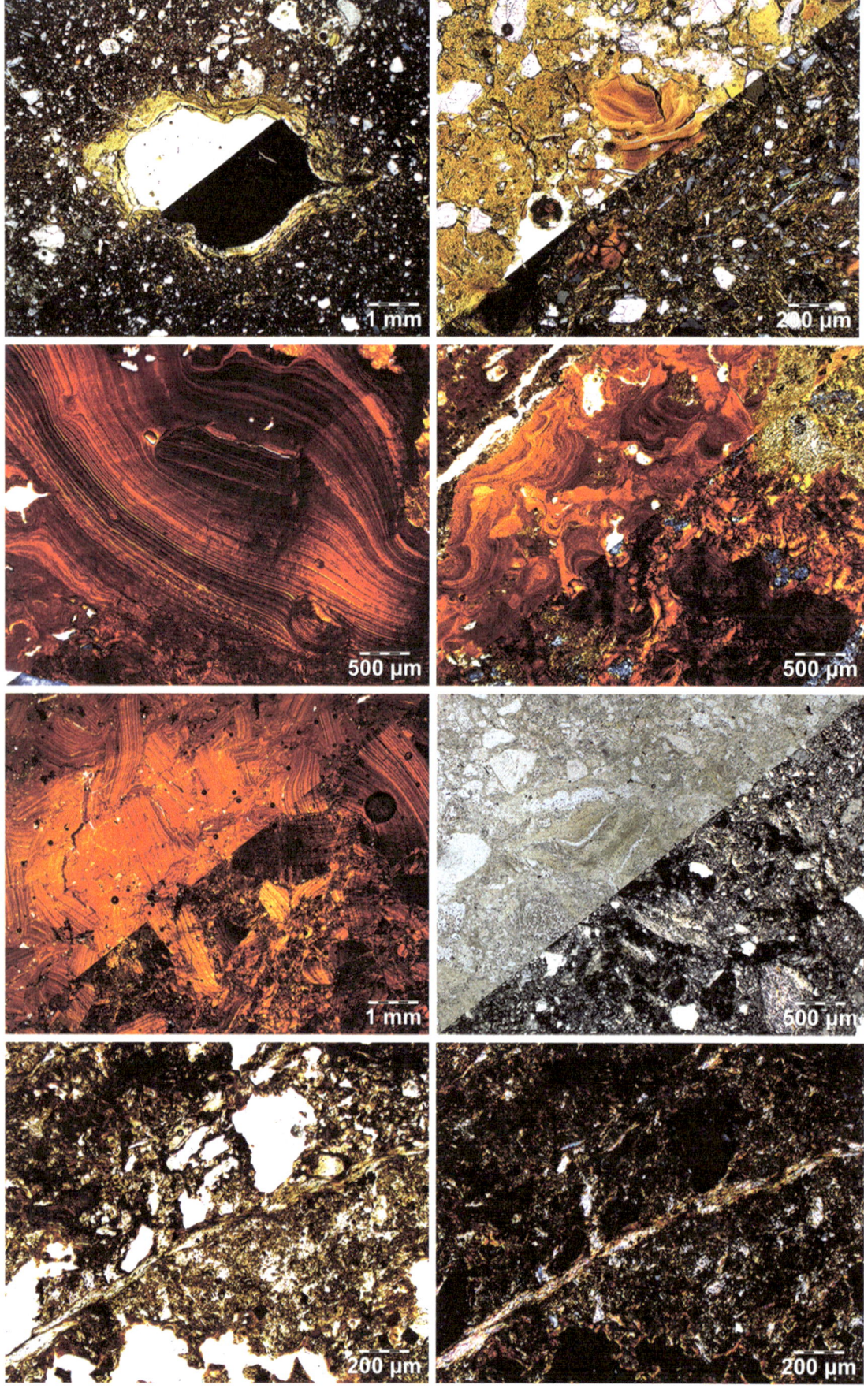

File 57: Coatings with Clays II

Coatings are defined as intrusive pedofeatures that coat natural surfaces of voids, grains, or aggregates inside soils (Stoops 2003, 2021). This section shows peculiar characteristics of coatings formed by clays associated to either reworking, massive deposition, or effects of waterlogging. Moreover, clay coatings must not be confused with clay neoformation in saprolite cracks, a case illustrated in the last two pictures.

Captions from upper left corner to lower right corner.

1. Typic clay coating around a cross-section of a channel. This situation is typical of a horizon affected by clay translocation. Chromic Luvisol, Apulia, Italy.

2. Fragments of laminated clay coatings isolated inside the groundmass. This type of feature indicates possible reworking of a former horizon affected by clay leaching. Such fragments have also been called "papulas" in the past (Brewer 1964). Reworked loess, Piedmont, Italy.

3. Very thick clay coating showing numerous alternating laminations. The red colour is probably due to the presence of nanocrystals of hematite. Clay translocation is so intense that almost the entire thin section is occupied by clay layers. Some of the contacts between laminations correspond to erosive surfaces. Chromic Luvisol developed in a cave, Sardinia, Italy.

4. Very thick and convoluted clay coatings formed by multiple laminations related to different translocation phases. Some of the contacts between laminations correspond to erosive surfaces. Paleosol, Ligurian coast, northern Italy.

5. Fragments of clay coatings formed during intense clay translocation (similar to 3.), which have been reworked and redeposited. Chromic Luvisol developed in a cave, Sardinia, Italy.

6. Clay intercalation observed in bleached zones of some waterlogged soils. According to Fedoroff and Courty (2012), such intercalations differ from typic clay coatings by their whitish grey colour, their absence of both sorting and lamination, and their medium to weak orientation. Soil in a rice chamber, Piedmont, Italy.

7.–8. In 7., example of the presence of clays inside a crack. Such a feature is not a clay coating, but a clay neoformation layer inside a saprolite. In 8., same view as 7. in XPL. Paleosol, Ligurian coast, northern Italy.

File 58: Micropans, Coarse Coatings, Cappings, and Crusts

Coatings are defined as intrusive pedofeatures that coat natural surfaces of voids, grains, or aggregates inside soils (Stoops 2003, 2021). In terms of morphologies, coatings can be subdivided into several types, such as the typic and crescent ones shown in "File 56" and "File 57". Micropans, crusts, and cappings are illustrated in this section. These types are often composed of material coarser than clay; consequently, this section also contains examples of coarse coatings.

Captions from upper left corner to lower right corner.

1.–2. A micropan is defined as "a thick sub-horizontal coating varying significantly in thickness" (Stoops 2003). In these two microphotographs, the groundmass is dominated by sandy quartz grains, emphasizing the presence of a textural pedofeature, the micropan being mainly composed of clay minerals (yellowish brown in PPL and undulated yellow with brown spots in XPL). In 2., close-up of the view shown in 1. highlighting the abundance of clay particles. Paleosol, Ligurian coast, northern Italy.

3. Compound layered coatings formed by alternating layers of clayey, silty, and sandy material. Clays with silt layers are sometimes described as dusty clay coatings, or generally speaking, coarse coatings. Paleosol, central Po plain, Italy.

4. Silt coating in a Luvic Stagnosol from Syria. Such coatings are formed by silt-sized grains, which can be fine, medium, or coarse.

5.–6. Cappings developed on large rock fragments (garnet mica schist; garnets are the black geometrical grains in XPL, whereas mica appear as yellowish to greenish bands; see p. 53). Cappings are composed of mainly coarse material, which form on top of free or embedded grains, and are often observed in Cryosols. Cryosol, western Alps, Lombardy, Italy.

7. Crusts are defined as thick coatings on the soil surface, whatever their grain-size distribution and nature (hypo- or quasicoating; see p. 117). The crust shown in this microphotograph developed on a silty groundmass material. It is formed by clay minerals as accentuated in the XPL view. Chromic Luvisol, Apulia region, Italy.

8. Coarse crust composed of silt, clay, and a few grains of sand on the top of a carbonate layer (grey in PPL and brownish grey in XPL), overlapping a yellowish–brownish silty–clayey groundmass (in XPL). Calcaric Cambisol, Apulia region, Italy.

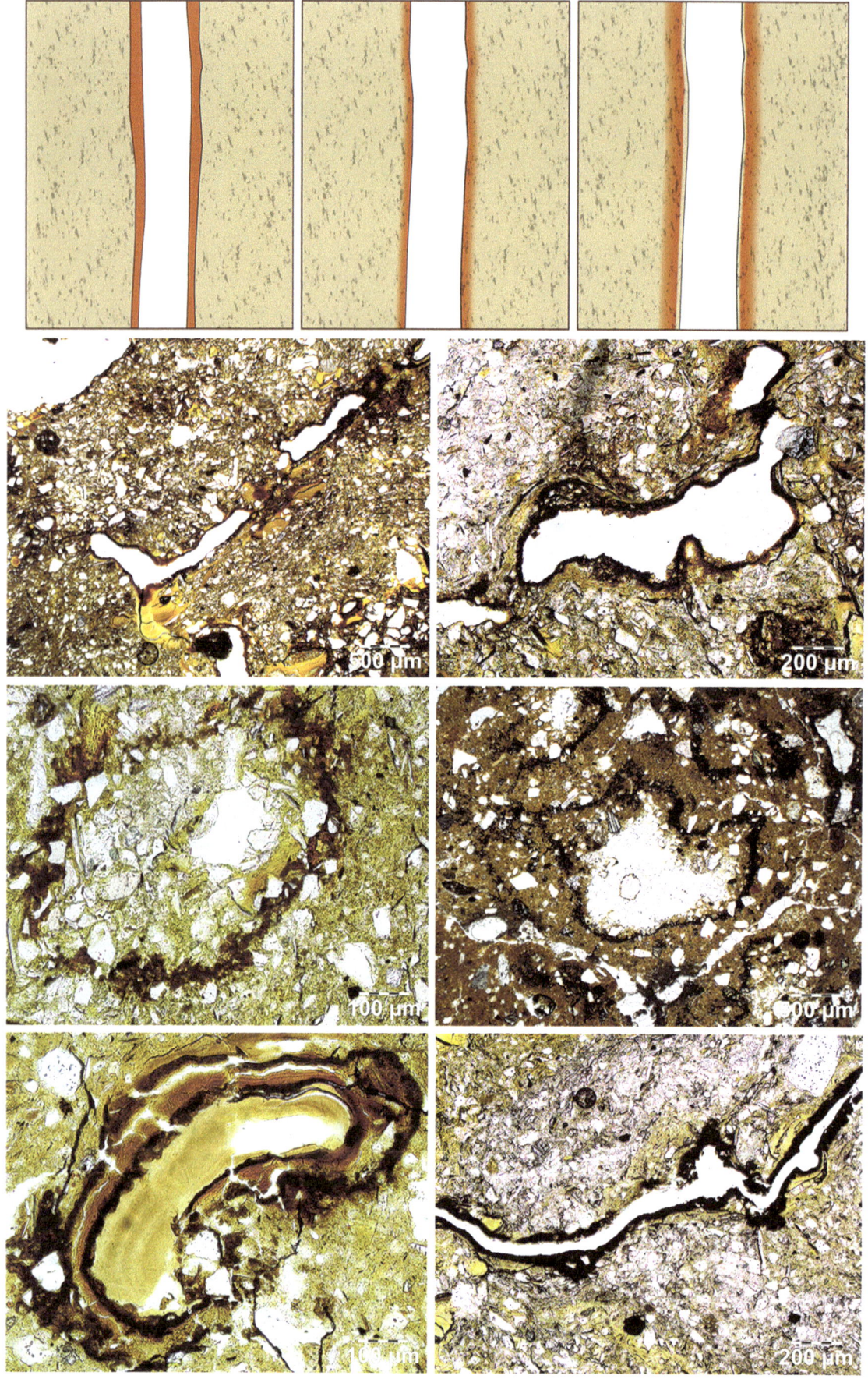

File 59: Hypocoatings and Quasicoatings: Amorphous

The difference between coatings, hypocoatings, and quasicoatings is the location of the material with respect to the internal soil surface. Hypocoatings refer to an accumulation of matter impregnating the soil groundmass directly adjoining the void edge (the internal soil surface). Quasicoatings are not in direct contact with the void border: there is a rim of soil groundmass material between the void and the quasicoating pedofeature. Because of their impregnative nature, hypocoatings and quasicoatings are mainly amorphous (this section) or crystalline ("File 60"). "Amorphous" refers to isotropic properties of iron and manganese oxyhydroxides, which mainly form these kinds of pedofeatures; according to Bullock et al. (1985), they can be regarded as amorphous or cryptocrystalline pedofeatures.

Captions start with the uppermost sketch, followed by the microphotographs from upper left corner to lower right corner.

1.   Sketch illustrating the geometrical relationship between coatings, hypocoatings, and quasicoatings (from left to right), the pore limits, and the soil groundmass.

2.–3.   Hypocoatings at two different magnifications made of an oxyhydroxide phase impregnating a loamy–sandy groundmass and associated with channel voids. Note the presence of fragments of clay coatings. PPL view, Stagnic Luvisol, Piedmont, Italy.

4.   Quasicoating of an oxyhydroxide phase impregnating a loamy–sandy groundmass. Note the presence of a groundmass rim between the void and the quasicoating. PPL view, Stagnic Cambisol, Apulia, Italy.

5.   Impregnating feature showing partial hypocoatings with quasicoatings resembling Liesegang rings. These hypocoatings with quasicoatings are composed of amorphous oxyhydroxides. PPL view, Stagnic Cambisol, Apulia, Italy.

6.   An example of both superimposed and juxtaposed compound pedofeatures: a first generation of dark quasicoating to hypocoating of amorphous material is followed by the deposition of a brownish clay coating, itself affected by a hypocoating of amorphous material. Finally, an incomplete infilling of yellow clays completes the succession of pedofeatures. See also "File 66". PPL view, Stagnic Luvisol, Apulia, Italy.

7.   Compound superimposed pedofeature consisting of a hypocoating of amorphous material affecting a clay coating along a longitudinal void. PPL view, Stagnic Luvisol, Piedmont, Italy.

File 60: Coatings and Hypocoatings: Crystalline

The difference between coatings and hypocoatings is the location of the material in respect to the internal soil surface. Coatings are intrusive pedofeatures that cover natural surfaces of voids, grains, or aggregates. Hypocoatings refer to an accumulation of matter impregnating the soil groundmass directly adjoining the void edge (the internal soil surface). This section includes various types of such crystalline pedofeatures (Bullock et al. 1985) that are related to void edges.

Captions from upper left corner to lower right corner.

1. Calcite coating around a void formed in a carbonate groundmass (with a micritic micromass). The calcite coating is formed by small-sized crystals (microsparite). Calcaric Leptosol, Jura Mountains, Switzerland.
2. Calcite coating around a void formed in a clayey groundmass. The calcite coating is formed by coarse crystals (sparite), probably originating from calcified root cells. A secondary clear calcite coating (microsparite) is juxtaposed to the coarse coating. Paleosol, central Sahara.
3. Calcite pendent developed on the lower surface of a sandstone grain. The pendent is made of banded and fibro-radial small grain-sized crystals. Cryosol, Spitsbergen, Svalbard Islands, Norway.
4. Calcite pendent forming multiple layers, starting with a fibrous coating in contact with the carbonate micromass (top), followed by a microsparitic band, and finally by large fans of fibrous calcite. Petric Calcisol, Madrid Basin, Spain.
5. Calcite hypocoating formed by small-sized crystals (micrite) impregnating a carbonate-rich micromass with some quartz grains. Bronze Age archaeological soil, Po Plain, Italy.
6. Juxtaposed crystalline pedofeatures: a hypocoating of microcrystalline calcite (micrite) is in contact with a microsparitic coating inside a root void. Bronze Age archaeological soil, Po Plain, Italy.
7. Hypocoating of phosphate around a cross-section of a channel in a carbonate-rich micromass. Medieval archaeological soil, Po Plain, Italy.
8. Transparent coating made of a cryptocrystalline siliceous compound (possibly silica) around quartz grains and carbonate-rich aggregates. The coatings are visible due to their high relief. PPL view, Duric Kastanozem, Manga, Niger.

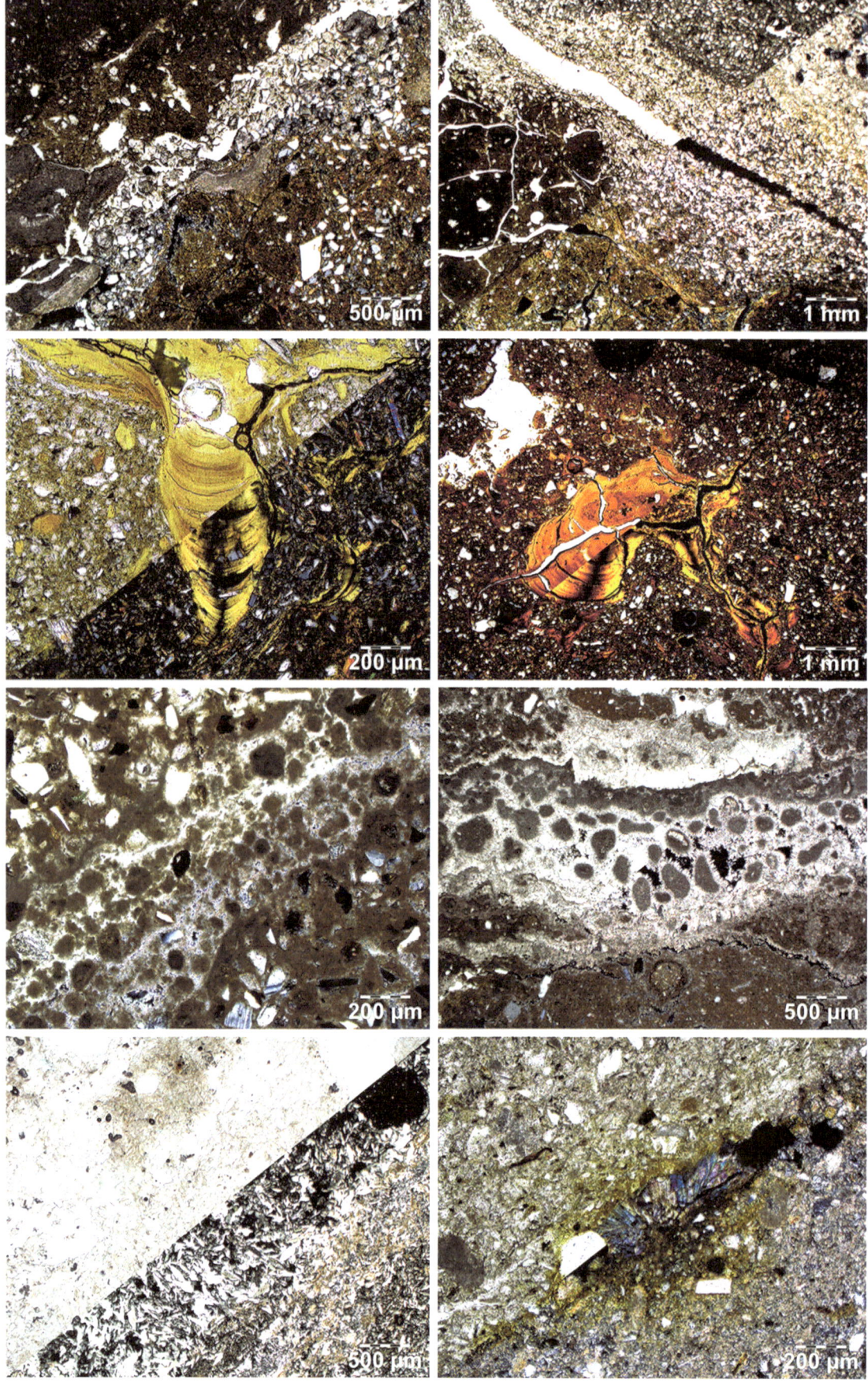

File 61: Mineral Infillings

> Infillings are formed by soil material or some fraction of it, which fills any void other than packing voids (Stoops 2003, 2021). These infillings are formed by mineral material originating from either biological (see "File 62") or physicochemical processes. This plate shows examples of mineral infillings, which can be coarse or fine and of different mineralogical natures. They are termed textural or crystalline pedofeatures according to their composition.

Captions from upper left corner to lower right corner.

1. A former planar void filled with coarse quartz grains, constituting a dense incomplete infilling. It can be considered as a textural pedofeature due to the grain-size distribution of the quartz grains. Agricultural layer of a Vertisol, Po Plain, Italy.
2. Well sorted and coarse infilling made by quartz grains forming a dense complete infilling, but which has undergone cracking during pedoturbation (central plane). In XPL, some slickenside features can be observed at the bottom of the microphotograph. Agricultural layer of a Vertisol, Po Plain, Italy.
3. Textural clay infilling in a silty groundmass. The clay nature of this dense incomplete infilling is accentuated by the colours and laminations in PPL as well as in XPL. Moreover, in the XPL view, the micromass is characterized by a circular striated b-fabric. Paleo-Luvisol, Piedmont, Italy.
4. Textural clay infilling in a silty to sandy groundmass. The clay nature of this infilling is accentuated by the colours and laminations in PPL as well as in XPL. The clays form a dense complete infilling, but which has undergone cracking due to pedoturbation. Paleo-Luvisol, Piedmont, Italy.
5. Infilling of micritic granular aggregates in a desiccation crack in a pre-existing carbonate-rich groundmass. These aggregates are associated with coarser calcite crystals (microsparite) as a secondary phase. Both features form a dense incomplete infilling. Petric Calcisol, Chobe Enclave, Botswana.
6. Large planar void developed in a carbonate-rich groundmass filled by a first generation of micritic aggregates, coated by microsparitic calcite cement deposited during phreatic events. These features form a dense incomplete infilling. In the upper part of the microphotograph, another planar void is filled by coarse-grained sparite forming a dense complete infilling. Petric Calcisol, Madrid Basin, Spain.
7. Large planar void filled with gypsum crystals, easily identified by their characteristic lenticular shape and their birefringence colours. These crystals form a loose discontinuous infilling. Gypsic Regosol, Swiss Alps.
8. Channel void in a carbonate-rich micromass filled with a crystal of vivianite, recognizable by its blue colours in PPL. This infilling is dense and incomplete. Reductive conditions are necessary for vivianite formation. Note the presence of phosphate hypocoatings. Medieval archaeological soil, Po Plain, Italy.

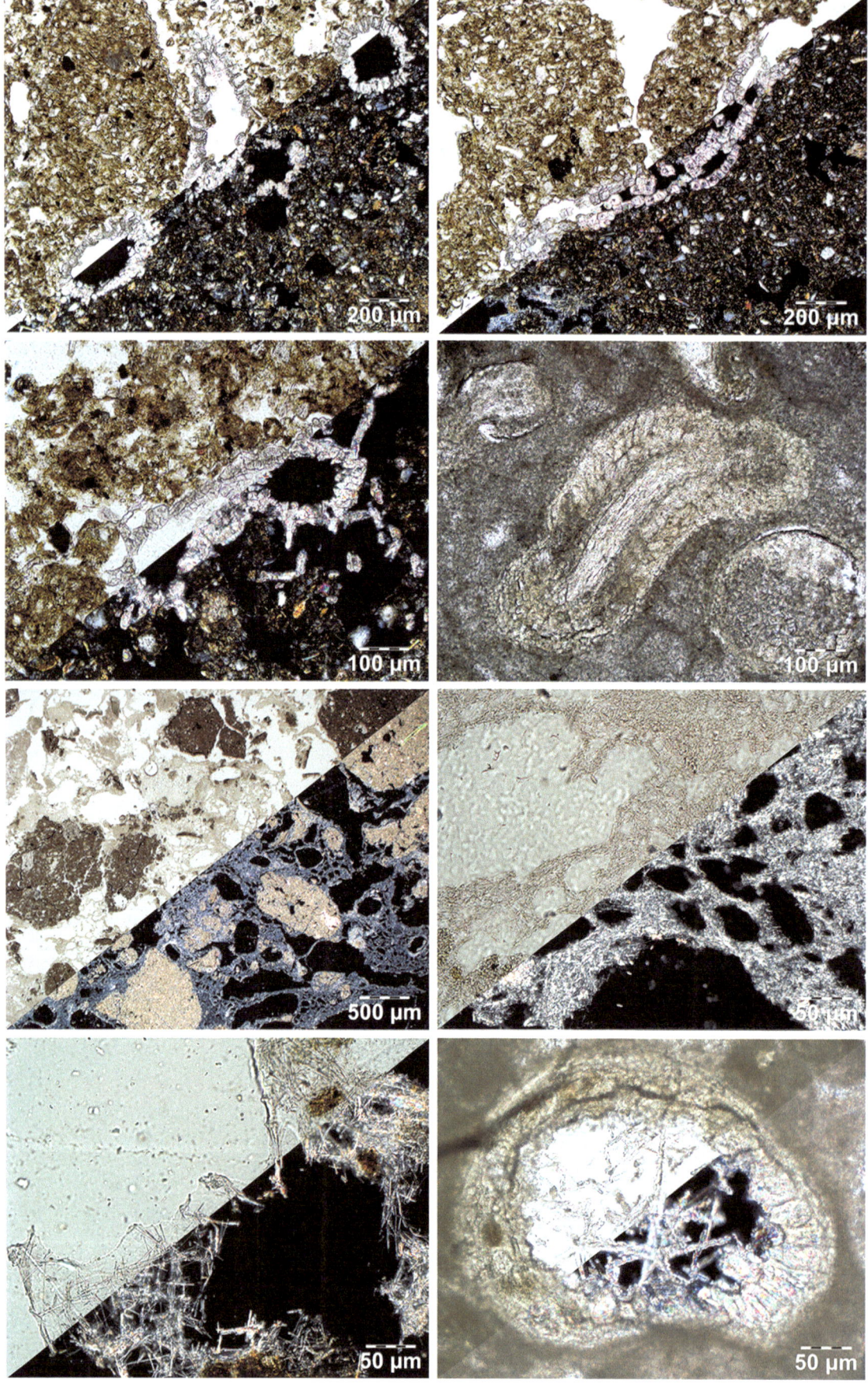

File 62: Mineral Infillings of Biological Origin

Infillings are formed by soil material or some fraction of it, which fills any void other than packing voids (Stoops 2003, 2021). These infillings are formed by mineral material originating from either biological (see "File 61") or physicochemical processes. This plate shows examples of biomineral infillings. They are termed crystalline pedofeatures according to their composition.

Captions from upper left corner to lower right corner.

1.–2. Dense incomplete infillings formed by calcified root cells (Becze-Deak et al. 1997; Durand et al. 2018) around cross-sections of (1.), and along (2.) channels in a siltic groundmass. In 1., most of the crystals originate from the biomineralization of epidermis cells, whereas in 2., the crystals were precipitated in cortex cells. Loess paleosol, Shaanxi, China.

3. Same type of feature as in 1. and 2. This infilling, due to calcified root cells, also preserved some biomineralized root hairs appearing as long styloidic calcite crystals overgrowing specific cells. Loess Plateau, northern central China.

4. Dense complete infilling in a carbonate-rich micromass, formed by the complete biomineralization of a root fragment, showing calcified cells from the epidermis, endodermis, and cortex of a rootlet. Petric Calcisol, Galilee, Israel.

5. A loose discontinuous infilling by secondary acicular crystals of calcite between micritic lithoclast of chalk. This type of infilling shows a convoluted fabric according to Rabenhorst and Wilding (1986). Petric Calcisol, Champagne, France.

6. Close-up of 5. showing the structure of the convoluted fabric made of acicular crystals of calcite. These crystals are not randomly organized but tend to form bundles. The origin of such features is related to fungal activity and biomineralization (Bindschedler et al. 2012, 2016). Petric Calcisol, Champagne, France.

7. Magnification of microphotograph 6. emphasizing the needle-fibre calcite nature of the acicular crystals (Durand et al. 2018). Petric Calcisol, Champagne, France.

8. A cross-section of a channel filled with a dense incomplete infilling made of calcified root cells (biomineralized epidermis cells), as well as a loose discontinuous infilling of needle-fibre calcite. In addition, the channel section is rimmed by a micrite hypocoating. Petric Calcisol, Galilee, Israel.

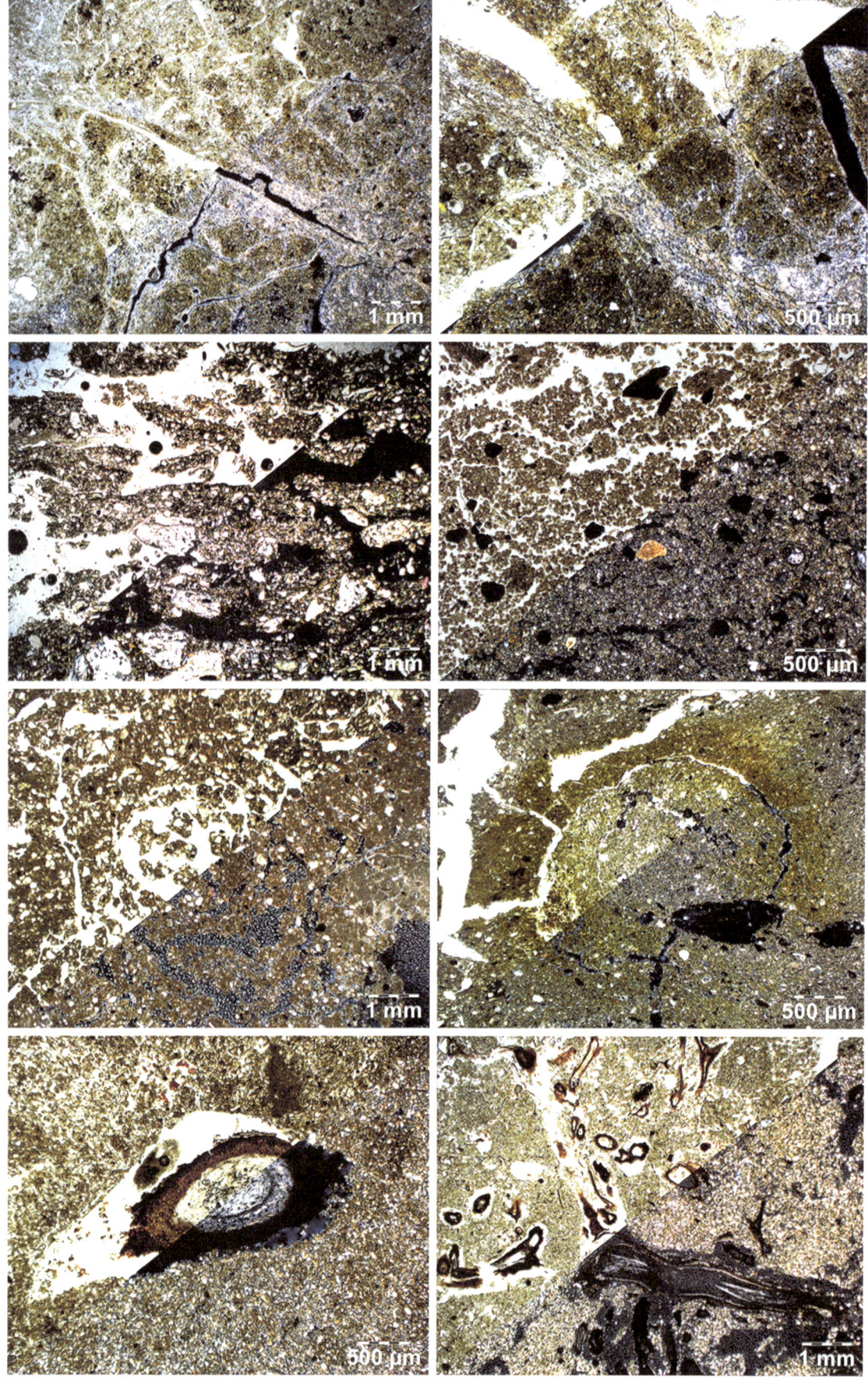

File 63: Pedoturbations

Pedoturbation is defined as any kind of physical mixing of soil material accomplished by the following mechanisms: shrinking and swelling of clays, freeze–thaw activity, and bioturbation by animals or plants. It is not because the soil material is mixed that it becomes homogenized (Schaetzl and Thompson 2015). However, even if some homogenization seems obvious at the macroscale, at the microscale, the effects of pedoturbation are always visible because of the imprint of the various processes. In addition, in Stoops (2003), most of the pedoturbations are included in fabric pedofeatures (belonging to matrix pedofeatures), which are "recognizable from the groundmass because of difference in fabric only".

Captions from upper left corner to lower right corner.

1.–2. Pedoturbation related to clay swelling and shrinking in a Vertisol. The brown and dark groundmass is separated by iso-oriented clay domains, i.e. slickenside, usually identifiable both in PPL (light greyish coloured) and XPL (first order birefringence colours), inducing a porostriated b-fabric. In 2., detail showing a clay domain at higher magnification. Vertisol, Po Plain, Italy.

3. Pedoturbation due to freeze–thaw activity and ice segregation forming lenses. Lenticular aggregates are covered by silt cappings. Both aggregates and cappings include fine sand grains. Cryosol, Alps, Italy.

4. Pedoturbation related to freeze–thaw activity in an organic-rich micro-granular horizon. A sub-lenticular aggregation is visible at the top of the image. Another level of aggregation, consisting of a network of zigzag planes, affects the groundmass in the lower part of the picture. The presence of micro-granular silty aggregates enhances the identification of frost pedoturbation. Paleosol, northern Apennines, Italy.

5. Pedoturbation made by animals. It can be considered as a passage feature, *senso* Stoops (2003). The groundmass is organized in a concentric manner but has been mechanically reworked and dispersed. Calcisol, Jura Mountains, Switzerland.

6. Pedoturbation due to animals. Note the yellow-coloured fabric hypocoating due to mechanical forces that have compacted the micromass. In this case, the groundmass has not been dispersed. Loess Plateau, northern central China.

7. Pedoturbation due to the presence of a rootlet (central part of the microphotograph in brown), which has created a large void and compacted its bottom part. Reworked loess deposit, northern Po Plain, Italy.

8. Pedoturbation due to a dense network of rootlets (brown rounded organic features in PPL) resulting in a disturbance of the groundmass. In XPL, a long channel is partly filled by the longitudinal cross-cut of a root. Fluvisol, Jura Mountains, Switzerland.

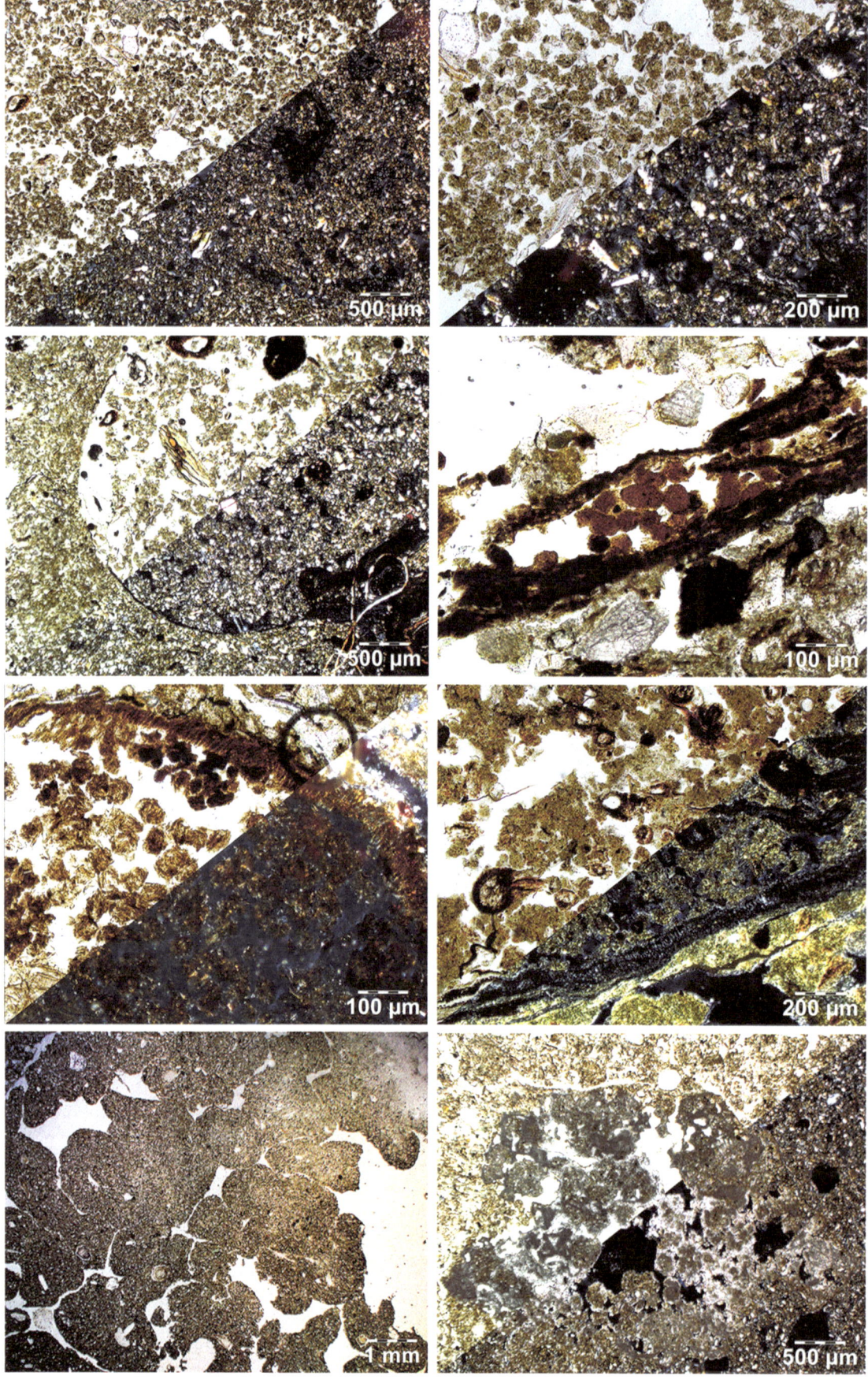

File 64: Faecal Pellets

Sometimes referred to as "excrements of the mesofauna" (Stoops 2003, 2021), faecal pellets (in the sedimentological sense of the word) do not only belong to a special type of pedofeature but can also create granular and/or vermicular microstructures (see "File 20" and "File 21"). They can form infillings as well. The faecal pellet's shape can refer to specific mesofaunas, as listed in Stoops (2003, 2021) and Bullock et al. (1985).

Captions from upper left corner to lower right corner.

1.–2. Layer extremely rich in faecal pellets forming a vermicular microstructure. Almost the entire horizon is composed of such excrements, resulting in a mixture of organic matter and clay-size mineral material. In 2., detail of microphotograph in 1. Paleosol, northern Apennines, Italy.

3. The channel tip in the central part of the microphotograph is partially infilled with faecal pellets and shows a sharp limit separating it from the groundmass. Fluvisol, Jura Mountains, Switzerland.

4. Ellipsoidal faecal pellets of Oribatid mites. The faecal pellets are weakly coalescent and loosely infill the hollow centre of a decaying plant fragment. Fluvisol, Swiss Plateau, Switzerland.

5. Faecal pellets found in a root, partially decayed by mesofaunas, which fed on plant organic matter and produced faecal pellets *in situ*. Faecal pellets are almost exclusively composed of organic matter. Calcisol, L'Isle-Adam, Paris Basin, France.

6. Faecal pellets found in a root, partially decayed by mesofauna, forming a mixture organic and mineral matter. The mesofaunas responsible for these faecal pellets are different from those which produced the faecal pellets in microphotograph 5. Cambisol, northern Apennines, Italy.

7. Large faecal pellets formed inside earthworm casts. These large faecal pellets are weakly coalescent and porous. It is sometimes difficult to distinguish them from the soil groundmass. However, these casts can be associated to calcitic biospheroids (see "File 38"). Cambisol, Swiss Plateau, Switzerland.

8. Coalescent faecal pellets of earthworms forming a partial infilling of a void. There is a clear difference in composition between the silty groundmass (yellowish brown in PPL) and the faecal pellets, which are impregnated by calcite (greyish colour in PPL). Loess Plateau, northern central China.

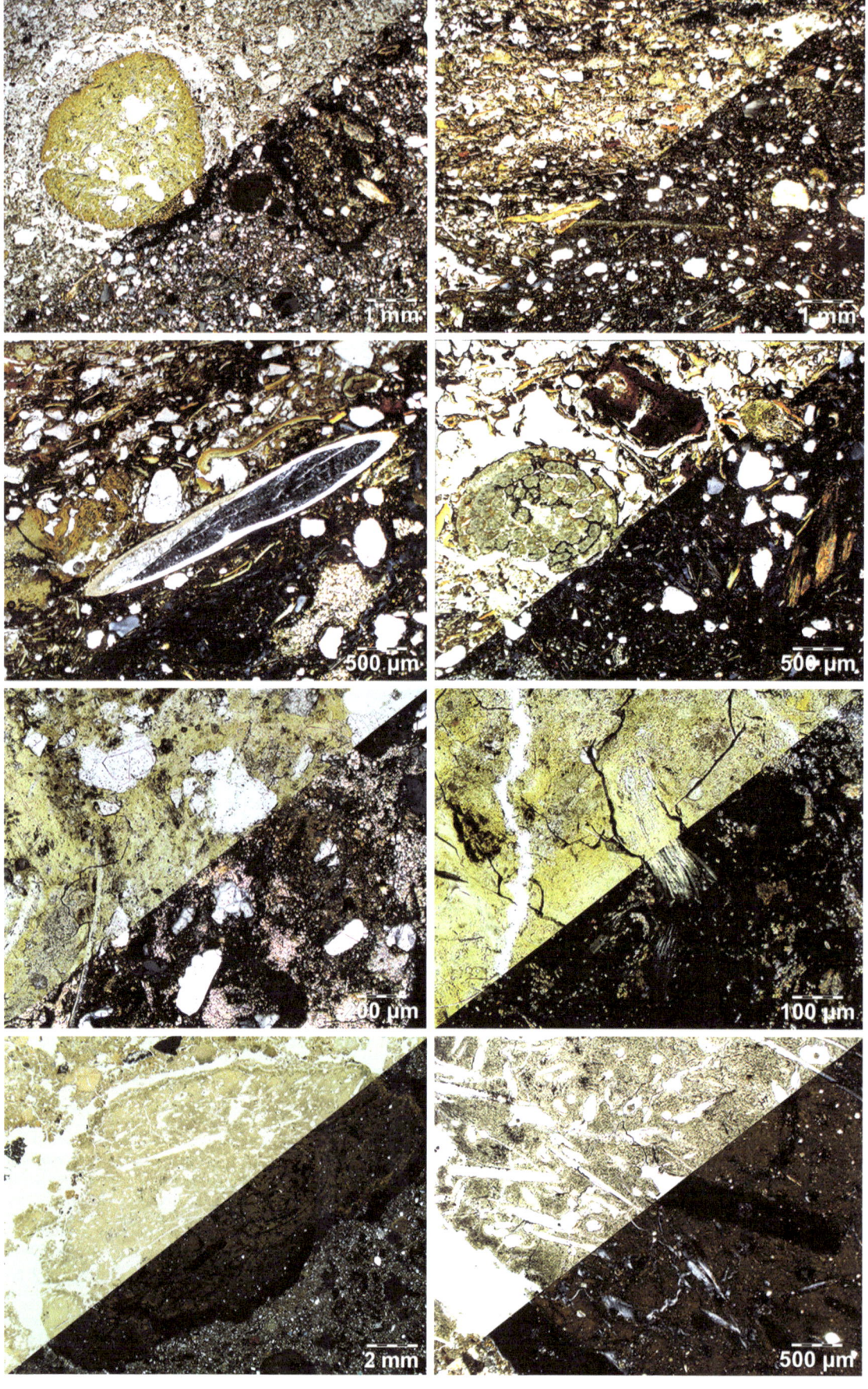

File 65: Dung and Vertebrate Excrements

Dung and excrements of large animals are mainly of interest to archaeological micromorphology (Nicosia and Stoops 2017), whereas faecal pellets can be observed in both natural soils and archaeological settings. Regarding excrements, they are characterized by different external shapes, colours, basic constituents, and internal fabrics, depending on the genera of the animals (Stoops 2003).

Captions from upper left corner to lower right corner.

1. Dung from a rock shelter in the Libyan Sahara showing the presence of undisturbed excrements of herbivores (capriovids), one in PPL and another in XPL. They are characterized by a sub-rounded mass with convoluted fabric, almost extinct in XPL (very low birefringence) due to the high content in organic matter. They might include opal phytoliths, faecal spherulites, oxalate druses, and crystals, depending on the diet of the animals (see "File 36").

2. Dung from a rock shelter in the Libyan Sahara showing a laminated fabric due to trampling by penned herbivores. Therefore, it is impossible to recognize any shape of excrement, although constituents of the faecal matter and biominerals (see "File 36") are still present but totally dispersed in the groundmass.

3.–4. Herbivore dung is composed of organic fragments at different stages of preservation (due to digestion) in both undisturbed and laminated deposits. In 3., the large oblong longitudinal shape is a seed in cross-section with long and thin plant fragments, whereas in 4., plant debris at various stages of preservation are observed along their radial section (see also "File 41").

5.–6. Bird excrements are yellow amorphous phosphatic masses, which can also contain uric acid spherulites (see "File 37"), as well as oxalate crystals (see "File 37"). In these thin sections, secondary precipitations of calcium carbonate form early diagenetic features, e.g. in 5., as high-order birefringence coloured clusters. Cave wall deposit, Libya, central Sahara.

7.–8. Carnivore excrement (*Hyaena* sp.). In 7., a large sub-rounded yellowish phosphatic mass containing bone fragments, as well as feather and hair imprints, as shown in detail in 8. Paglicci Cave, Apulia, Italy.

File 66: Composite Pedogenic Features

Pedofeatures made up of several parts or elements (i.e. composite) encompass compound and complex pedofeatures, as defined by Stoops (2003). They consist of a mixture of two or more pedofeatures resulting from different pedogenic processes, either contemporaneous or successive. If each single pedofeature lies side by side, they are defined as juxtaposed, whereas if they overlap, pervade, or affect one another, they are considered to be superimposed.

Captions from upper left corner to lower right corner.

1. Compound pedofeature constituted by three generations of clay coatings, juxtaposed to one another. The three coatings are distinguishable by their respective colours, from orange, at the bottom, to bright, and finally pale yellow. These juxtaposed clay coatings form a clay infilling inside a channel. Paleo-Luvisol, Piedmont region, Italy.
2. Succession of three different pedofeatures around a channel in a loamy groundmass. First, an amorphous hypocoating (orange with coarse material) impregnated the micromass. A red and laminated clay coating formed the second step. Finally, a light yellow and non-laminated clay coating ended the sequence. The three pedofeatures are juxtaposed to one another. Paleosol, Liguria, Italy.
3. Reworked fragment of two juxtaposed clay coatings clearly distinguished in PPL due to their different colours (light and dark orange). Paleosol, Liguria, Italy.
4. Reworked fragment of a clay coating surrounded by microsparitic nodules. These two juxtaposed pedofeatures refer to two distinct and successive pedogenic processes. Paleosol in a doline infilling, Syria.
5. Dark amorphous iron-bearing nodule partially overlapped by a yellow clay coating. In this case, the relationship between the two pedofeatures refers to juxtaposition. Paleo-Luvisol, Piedmont, Italy.
6. Amorphous iron oxyhydroxide hypocoatings superimposed on large sparitic crystals of a calcium carbonate nodule. Paleosol in a doline infilling, Syria.
7. The groundmass has been first depleted (colour gradient from the side to the centre). A new clay infilling in its central part (bright orange) superimposes this depletion pedofeature. Paleosol, central Po Plain, Italy.
8. Oxyhydroxide hypocoating superimposes a phosphate coating (yellow in PPL and extinct in XPL). Medieval archaeological site from northern Italy.

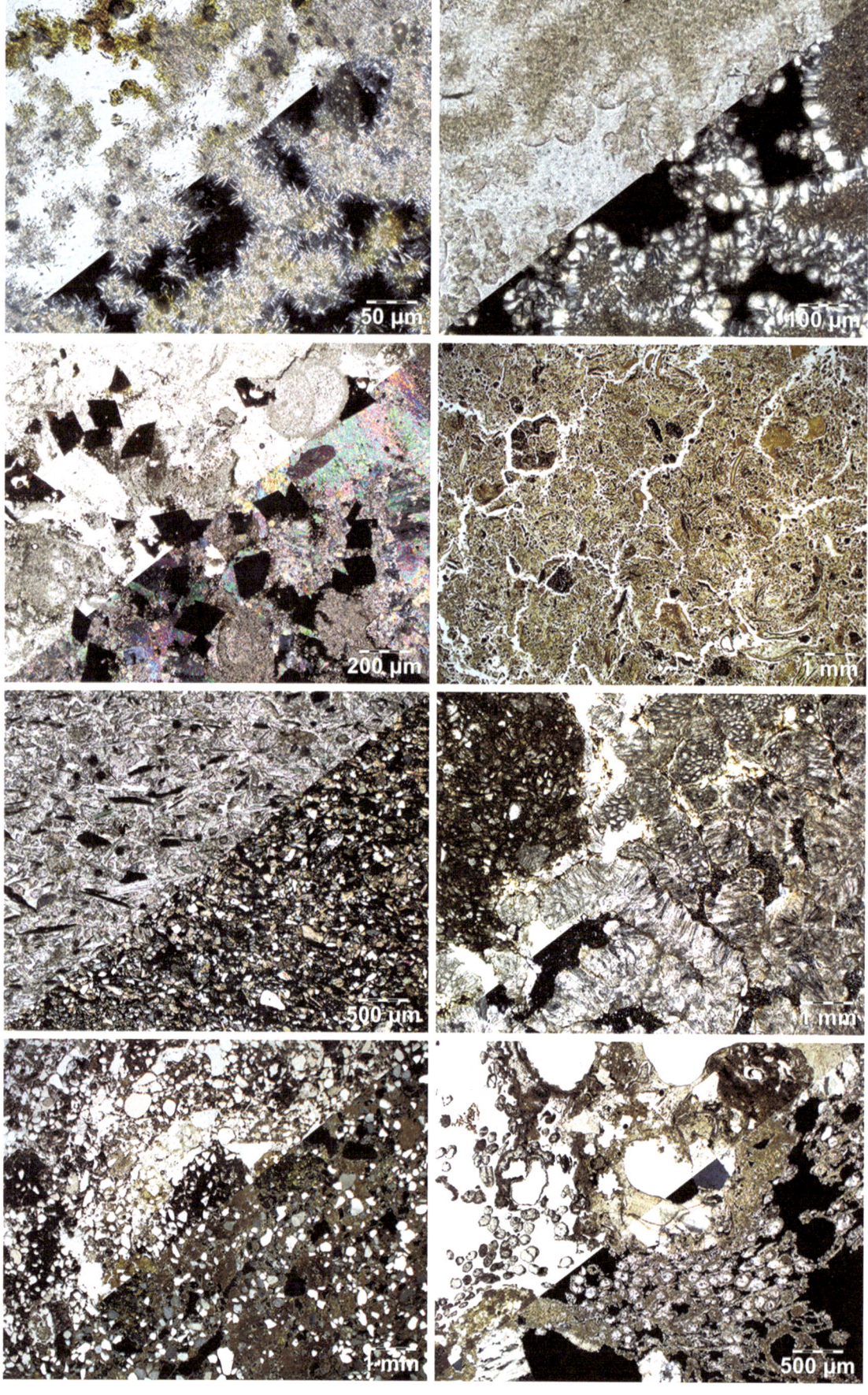

File 67: Uncommon Features

The list of uncommon features encountered in soil thin sections could be very lengthy. In this section, only eight have been selected and concern specific minerals, forms of organic matter, and finally some micromorphological effects produced by termites and tropical trees. The examples chosen are sometimes confused with other features.

Captions from upper left corner to lower right corner.

1. Secondary calcium carbonate forming random crystal intergrowths of needles. The needle shape does not refer to needle-fibre calcite but to acicular crystals of aragonite. The presence of aragonite is due to the high concentrations of ions, such as Na^+ and Mg^{2+} in the soil solution. Vertic Solonetz, Dead Sea, Israel.
2. Crystalline coatings of sodium silicate forming a palisade fabric, i.e. juxtaposed fans of magadiite (Sebag et al. 2001). These minerals form under high-alkaline conditions in brines. Chernozem, Lake Chad, Niger.
3. Soil developed on a marine limestone in which dark crystal intergrowths of iron oxyhydroxides formed. These crystal infillings have the rhombohedral shape of former dolomite crystals, which have been dissolved during an early stage of pedogenesis. Calcic Cambisol, Provence, France.
4. Sapric horizon from a peat bog soil in which less than one-sixth of the groundmass is recognizable as original plant material (Chesworth 2008). Only a few fragments can be ascribed to cells and fibres of plants; most of the thin section is composed of an undefined and brownish material, forming a micromass affected by zigzag planes, delimiting aggregates. Histosol, Jura Mountains, Switzerland.
5. River deposit with a coarse monic c/f related distribution. The long and black shards in PPL and extinct in XPL are fragments of graphite and not present-day humified organic matter. Fluvisol, Rhone Valley, Switzerland.
6. A specific pedofeature encountered in paleosols, mostly from the Tertiary, called *Microcodium*. These crystal intergrowths are fossil features and must not be confused with present-day or Quaternary calcified root cells. Their shape resembles a corn cob with a single layer of elongated cells around a hollow or infilled (recrystallized) central axis. Paleosol, Corbières Massif, France.
7. Perturbed soil reworked by termites. The groundmass includes quartz grains and the micromass is a mixture of calcite and diatomite, the former being brownish grey coloured and the latter extinct in XPL. Aggregates are also present. Kastanozem, Chobe Enclave, Botswana.
8. Biomineralization of iroko wood cells forming calcified cells. The micromass is micritic with some large sparitic crystal intergrowths. Such features are associated to the oxalate–carbonate pathway (Cailleau et al. 2005). Calcite layer in a Ferralsol, Biga district, Ivory Coast.

Chapter 5

Pedofeatures Associated to Soil Processes

File 68: Pedofeatures and Soil Processes

As stipulated by G. Stoops, "the aim of micropedology is to contribute to solving problems related to the genesis, classification and management of soils, including soil characterization in palaeopedology and archaeology. The interpretation of features observed in thin sections is the most important part of this type of research, based on an objective detailed analysis and description" (Stoops et al. 2018). To answer such questions, two major books contributed to the comparative knowledge necessary to tackle this objective: the first one was published in 1985 and used micromorphology to distinguish between different classes of soils (Douglas and Thompson 1985); the second one is an extensive guide of more than 1000 pages to the interpretation of micromorphological features encountered in thin sections of soil (Stoops et al. 2018). The aim of this Atlas is neither to be a substitution for these books nor a way to enter directly into the interpretation of soil genesis and classification. Nonetheless, this chapter presents the imprints of major soil processes that can be easily deduced from specific features observed in thin sections. These processes involve the dynamics of (a) clay, both translocation and swelling, (b) water, such as waterlogging, evaporation, and its role as ice and frost, (c) carbonate, gypsum, and iron oxyhydroxides, and finally (d) biogeochemical reactions within the solum.

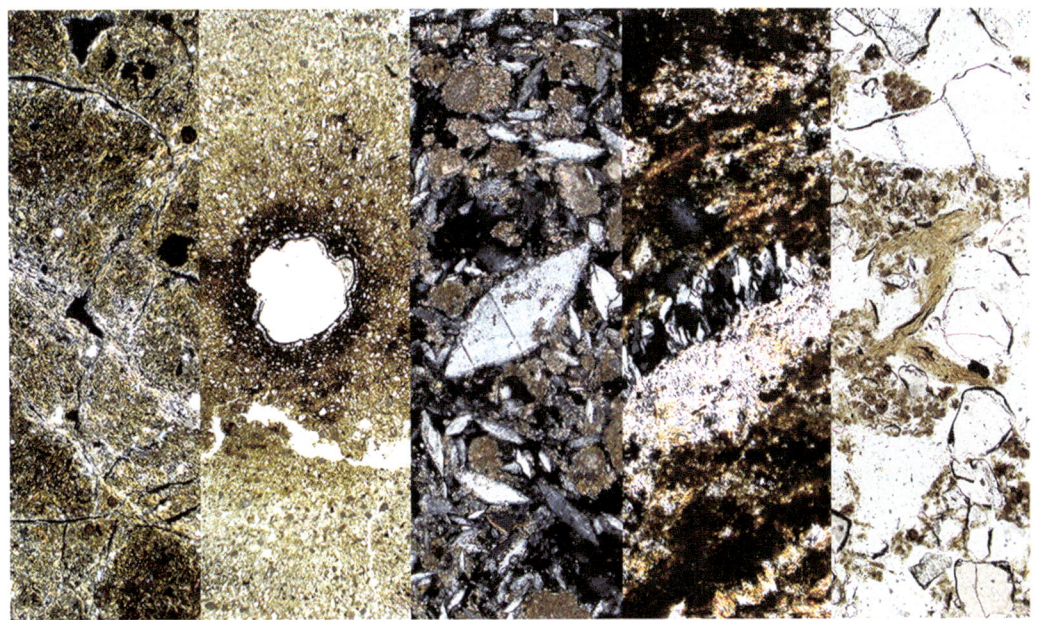

Examples of micromorphological expressions due to specific pedogenic processes: vertic material (XPL), a hydromorphic feature (PPL), precipitation of sulphate in an arid environment (XPL), clay neoformation (XPL), the influence of podzolization (PPL).

E. P. Verrecchia, L. Trombino, *A Visual Atlas for Soil Micromorphologists*, https://doi.org/10.1007/978-3-030-67806-7_5

135

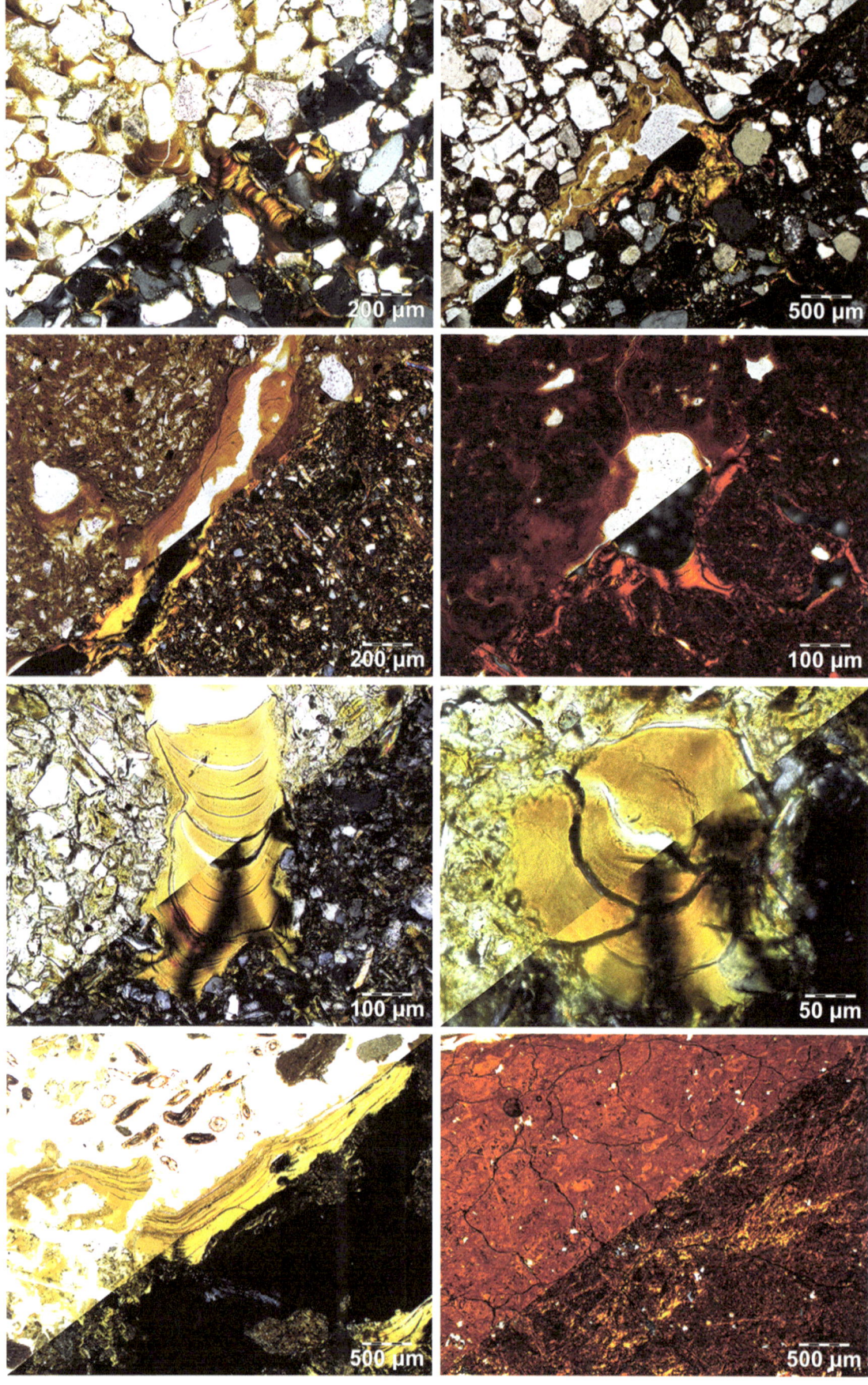

File 69: Clay Dynamics I: Translocation

Translocation refers to the mechanical process of displacing clays (or other material) in their dispersed state, i.e. in the form of isolated particles (Schaetzl and Thompson 2015). This process is related to vertical or lateral water movements in the soil profile. It is one of the first observed processes in thin section during the last century. This process can affect a large variety of soils, changing the associated features (textural pedofeatures; see "File 56", "File 57", "File 58" and "File 61") in terms of grain size and mineralogy. The soil porosity plays a major role in translocation, i.e. in mobilization and sedimentation. In addition, the soil chemistry can also influence the dispersed state of the mineral particles. Finally, climate (including seasonality) has a distinct impact on translocation, as free percolating water is needed to displace the particles.

Captions from upper left corner to lower right corner.

1. Clay translocation forming coatings and infillings in a sandy soil. It is important to consider the type of void affected by the translocation, in this case, packing voids. During the process, the clay translocation modified the c/f related distribution from coarse monic to close porphyric, which is shown in the microphotograph. Cambisol, Paris Basin, France.

2. Clay translocation forming coatings in a soil with two grain-size modes (quartz grains and a clayey micromass). The voids affected by the translocation are mainly connecting voids, i.e. channels and vughs. Paleosol, Isle of Elba, Italy.

3. Clay translocation forming coatings in a silty soil. The voids affected by the translocation are connecting voids, i.e. channels. As the process continues with time, all the clay coatings will become infillings. Paleo-Luvisol, Piedmont, Italy.

4. Clay translocation in a clayey soil. The distinction between the pedofeatures and the micromass can be challenging due to the low contrast between the two phases. However, the use of XPL can facilitate this distinction, even if sometimes it remains difficult to distinguish between clay coatings and a striated b-fabric. Chromic Luvisol, Apulia, Italy.

5.–6. Clay translocation observed in two different directions of a cross-section: vertical (5.—laminations clearly appear) and transversal (6.—concentric fabric). Paleo-Luvisol, Piedmont, Italy.

7. After a translocation, clay coatings can be dismantled by bioturbation, which can result in fragmentation without deformation. Cambisol, Apennines, Italy.

8. Argilliturbation or shrinking and swelling of clays (Schaetzl and Thompson 2015) resulting from the integration of pedofeatures due to clay translocation into, and forming, a striated groundmass. Whatever the type of original material, a massive translocation of clays can generate argilliturbation. Paleosol, Lombardy, Italy.

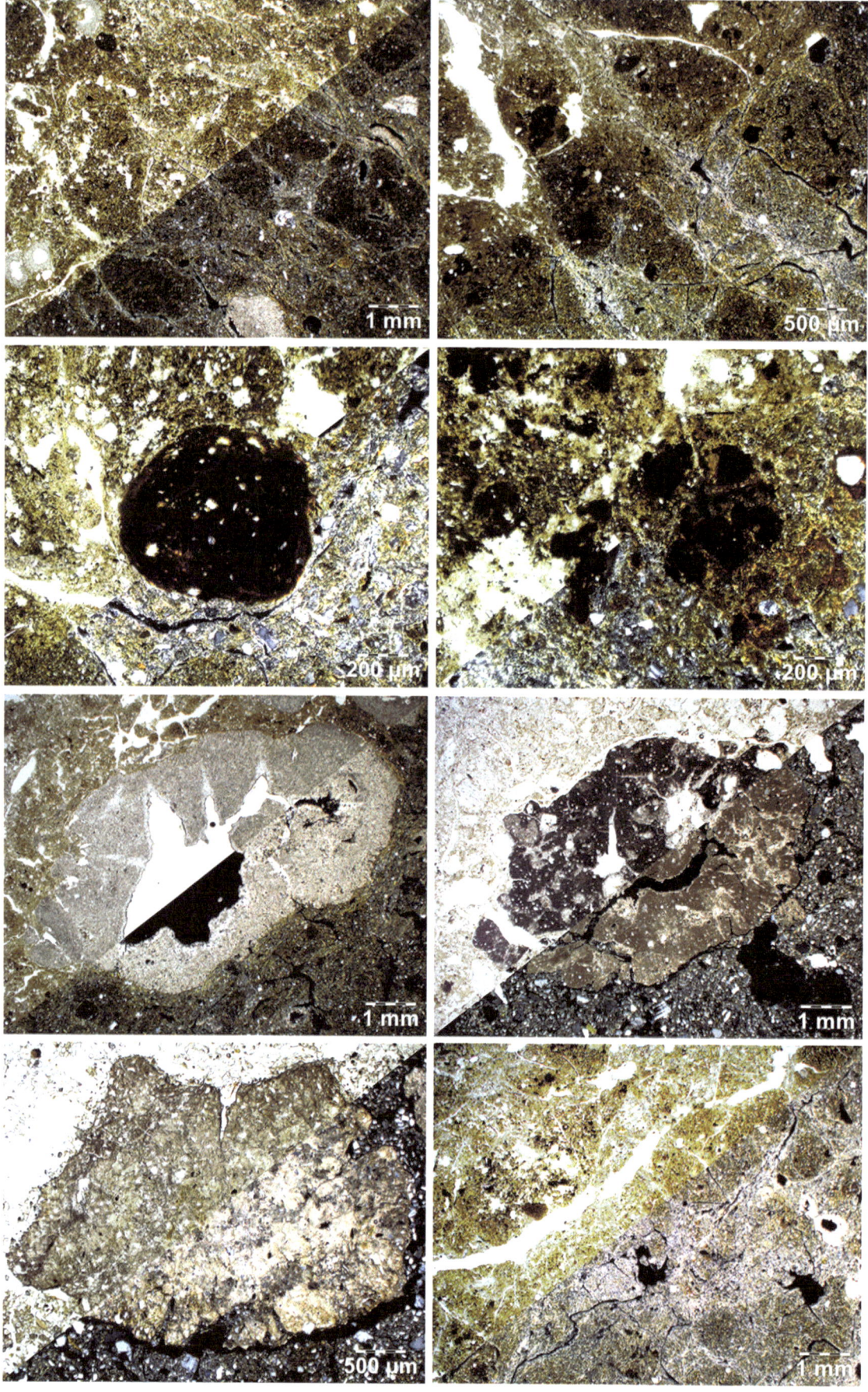

File 70: Clay Dynamics II: Swelling

> Some soils are characterized by a parent material containing a high proportion of TOT clays. Some clays have the property to shrink and swell depending on soil moisture saturation, leading to a pedogenetic process known as vertisolization and a specific class of soils, i.e. Vertisols (Duchaufour 1977; Legros 2012). Vertisols exhibit distinct characteristics, which are described in this section.

Captions from upper left corner to lower right corner.

1. General view of a Vertisol horizon in which the b-fabric is striated, a characteristic induced by shrinking and swelling of clays (see "File 69", microphotograph 8.). Vertisol, Po Plain, Italy.
2. Close-up of 1. showing slickensides (NW to SE bright light striation in XPL): these features are due to the reorientation of clays related to shrinking and swelling. For this reason, they were called "stress coatings" by Brewer (1964). Vertisol, Po Plain, Italy.
3. Iron-bearing nodule with a sharp boundary. This boundary is not due to its allochtonous origin (see "File 51") but to seasonal reworking related to shrinking and swelling of the clay micromass. Vertisol, Po Plain, Italy.
4. Iron-bearing aggregate nodule formed by the aggregation of small nodules. They are common pedofeatures of Vertisols (Stoops 2003). Vertisol, Po Plain, Italy.
5. Septaric morphology is frequent in Vertisols; in this case, it is formed by microsparitic crystals, without any evidence of groundmass impregnation: the origin of such intrusive pedofeatures remains unexplained, although a biogenically mediated process is likely. Paleo-Vertisol, Far North district, Cameroon.
6. Septaric micritic nodule showing two generations of calcitic crystals: the main fabric of the nodule is fine grained (micrite in dark brown). A second generation of cement fills some of the nodule pores with a microsparitic infilling (light grey). The most common calcitic nodules in Vertisols are micritic to microsparitic. Paleo-Vertisol, Far North district, Cameroon.
7. Nodule found in a Vertisol with a fan-like calcitic fabric. Such fabrics in nodules are uncommon but have also been observed in the geological record and described as cone-in-cone structures in pedogenic nodules (Freytet et al. 1992). Paleo-Vertisol, Far North district, Cameroon.
8. If a Vertisol is particularly enriched in a calcium carbonate phase, not only can calcitic nodules be precipitated but small calcitic crystals can also invade the micromass, forming a crystallitic b-fabric. This crystallitic b-fabric potentially obliterates the striated b-fabric described in microphotograph 1 (this section). Vertisol, Po Plain, Italy.

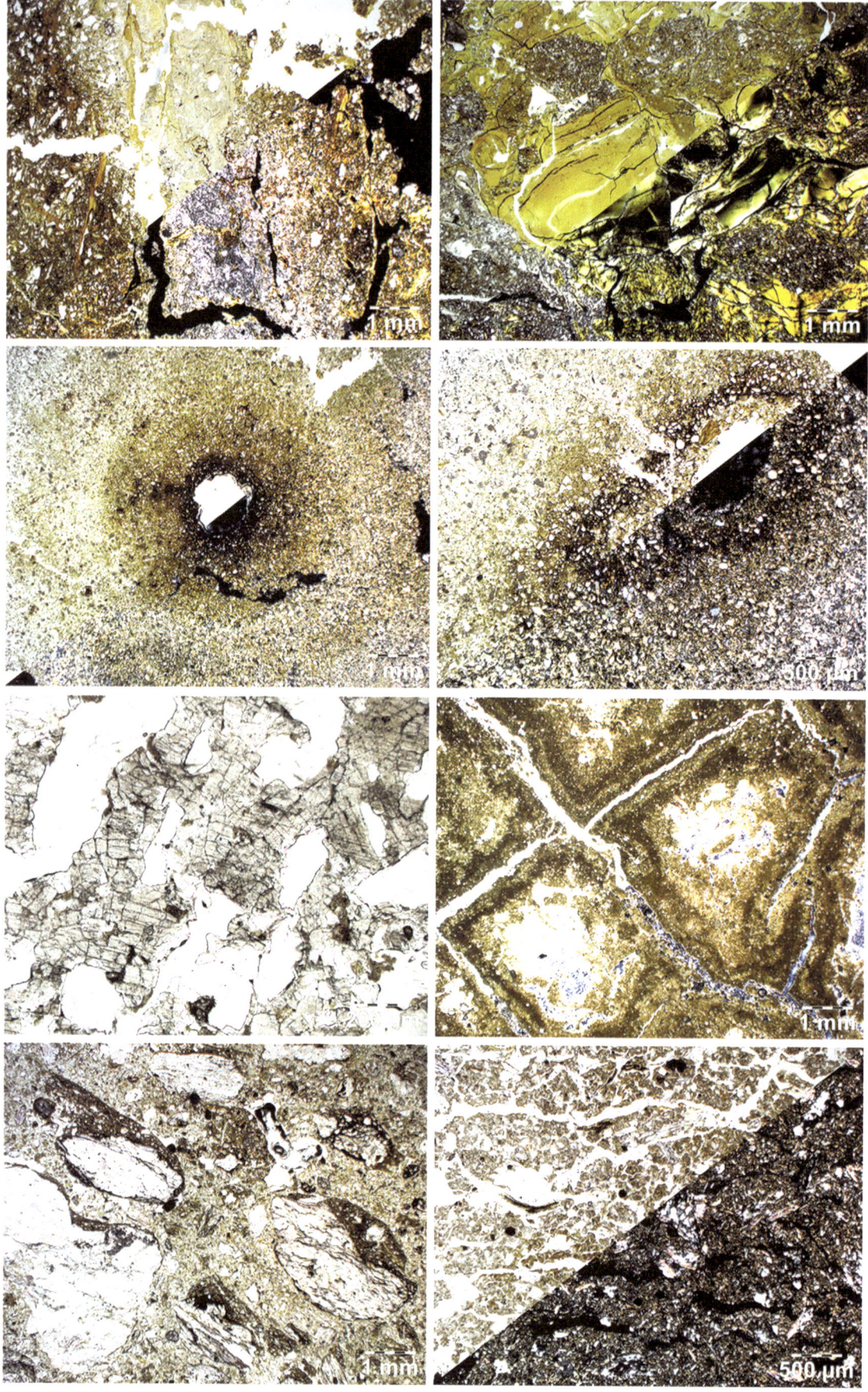

File 71: Water Dynamics

Water is the main force in pedogenesis. It influences soil dynamics in many ways. In this section, three types of water-driven processes are considered: waterlogging, evaporation, and water in freezing/thawing environments.

Captions from upper left corner to lower right corner.

1. In waterlogged soils, the presence of water precludes the infiltration of oxygen, triggering redox processes at the origin of hydromorphism (Schaetzl and Thompson 2015). In this microphotograph, iron has been reduced and removed from the micromass and the pedofeatures (coatings). This results in depletion pedofeatures (see "File 55"), which are lighter coloured compared to the surrounding material. Stagnic Luvisol, Piedmont, Italy.

2. In hydromorphic soils, it is common to find clay coatings and infillings. They usually display a whitish-grey colour and a quasi-absence of lamination and sorting (see also "File 57"). Both characteristics are due to waterlogging: the colour to the reduction of iron and its removal, and the internal structure to in situ redeposition. Soil in rice chamber, Piedmont, Italy.

3.–4. Common pedofeatures in hydromorphic soils are hypocoatings (3.) and quasicoatings (4.) developed around voids and related to iron redox dynamics (see "File 59"). The microphotographs show the influence of rootlet channels, in which oxygen could circulate, on iron distribution. Therefore, there is a gradient of amorphous iron-oxyhydroxide concentrations from the groundmass towards the void, mimicking the oxygen gradient. Fluvisols, Jura Mountains, Switzerland.

5. Precipitation of coalescent cubic crystals of halite in a Solonchak horizon due to intense evaporation of brackish groundwater. Such chloride-rich deposits correspond to an evaporitic sequence, sometimes ending with bromides. PPL view, Dead Sea shore, Israel.

6. Horizontal cross-section in a Solonetz horizon. Desiccation planes separate prismatic aggregates constituted by a core containing ghosts of halite crystals with an outer clay layer (dark brown), forming a quasicoating. Evaporation of the brackish water at the bottom of the soil induces both precipitation of chlorides and formation of large planes in a muddy micromass. Dead Sea shore, Israel.

7. In soils undergoing freezing/thawing phases, coarse cappings develop on large grains (see also "File 58"). Because of the reworking related to frost, some cappings can be detached from their supporting grains and incorporated into the groundmass or can be reoriented forming downturned cappings (van Vliet-Lanoë and Fox 2018). PPL view, Cryosol, Italian Alps.

8. Frost can induce lenticular microstructures (see "File 21") with plate aggregates (see "File 15"), resulting from the formation of ice lenses (van Vliet-Lanoë and Fox 2018). Cryosol, Italian Alps.

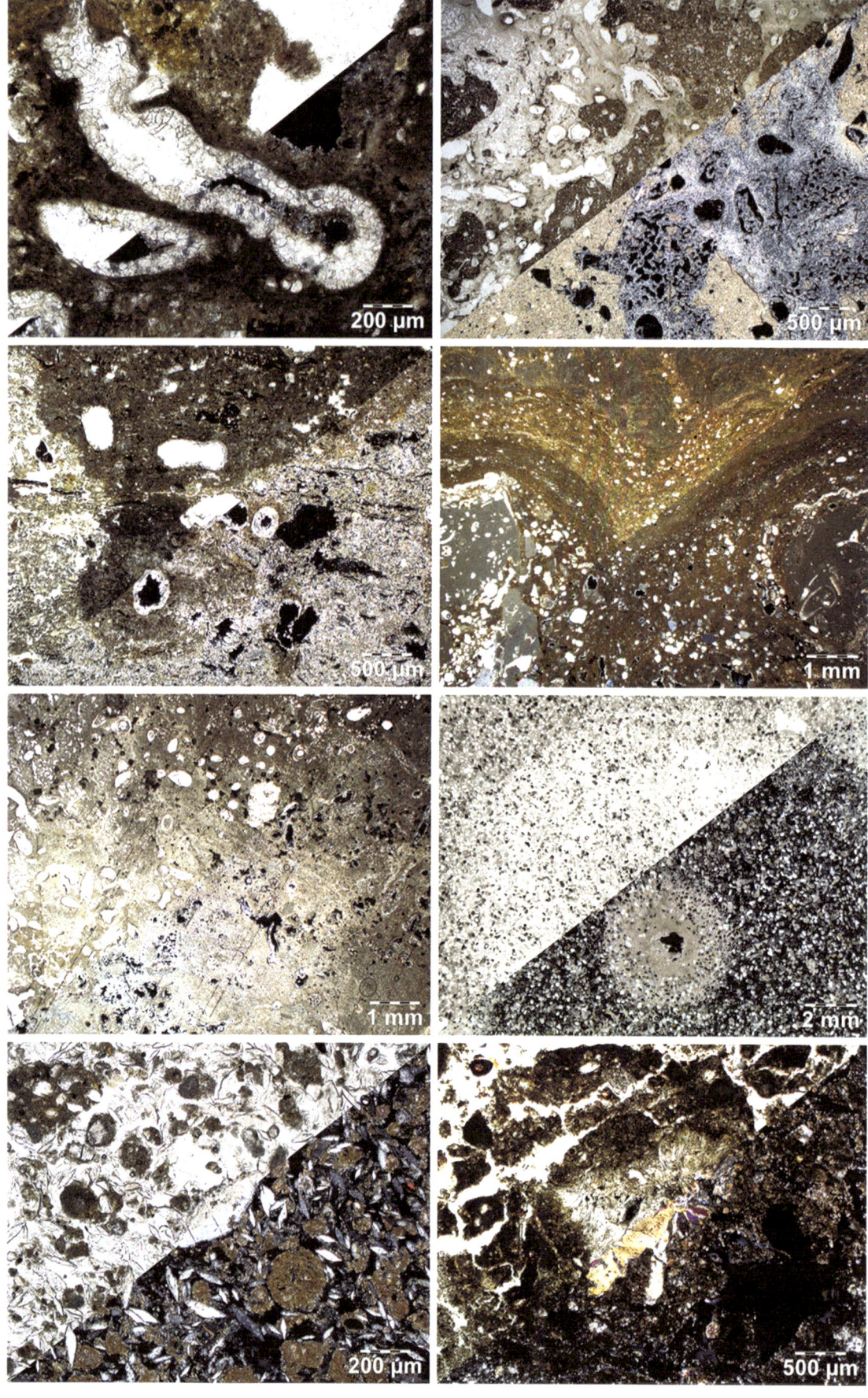

File 72: Carbonate and Gypsum Dynamics

> Pedogenic carbonate-rich soils must not be confused with *calcrete*, although this confusion is widespread in the literature. Calcium carbonate redistributions in soils can occur under various climates and are not limited to the semiarid and arid zones. Indeed, large redistributions of $CaCO_3$ are caused by organisms, i.e. bacteria, fungi, roots, and animals (Durand et al. 2018). Many pedogenic carbonates result from the incorporation of soil CO_2 (Hasinger et al. 2015) and are not the consequence of a simple dissolution and reprecipitation from a carbonate parent material. Gypsum dynamics necessitate dissolution and reprecipitation of sulphate through evaporation processes. There are some exceptions related to anthropogenic influences, such as acid rain.

Captions from upper left corner to lower right corner.

1. Calcium carbonate redistribution through the action of roots. Calcified root cells (see "File 42" and "File 62") associated to a thick micritic hypocoating have been precipitated during root activity. Fluvisol, Dorigny plateau, Switzerland.
2. Mineral infilling by needle-fibre calcite between chalk fragments (see "File 62"). Pores originate from rootlets, whereas needle-fibre calcite is precipitated by fungal filaments. Calcisol, Champagne, France.
3. Impregnation of the groundmass by secondary calcium carbonate, inducing a crystallitic b-fabric (see "File 45"). Note the presence of channels and calcified root cells. This soil is subject to water-table fluctuations. Bronze Age archaeological site, Po Plain, Italy.
4. Laminar horizon associated to a petrocalcic horizon. The laminated feature (upper half of the microphotograph) is not related to a *per descensum* process but rather to the superficial lithification of a biological mat (see "File 50"). Petric Calcisol, Negev Desert, Israel.
5. The lower part of the microphotograph has been depleted in calcium carbonate. This process is due to both dissolution during the wet season and respiration of rootlet mats. Note the presence of a convoluted fabric (see "File 62"). Calcisol, Galilee, Israel.
6. Large depletion rim, concentric to a root channel (extinct in XPL). Calcium carbonate has been redistributed from the soil groundmass (cemented by microsparite, upper part of the microphotograph) to the root channel, forming a grey micritic hypocoating (in XPL). Calcisol, Sharon, Israel.
7. Crystal intergrowths of gypsum obliterating the carbonate groundmass and disrupting aggregates. Gypsisol, Negev, Israel.
8. Recrystallization of secondary sulphate (anhydrite) after dissolution of primary gypsum. Such crystal intergrowth possibly indicates very dry local conditions. Gypsiric Leptosol, Valais, Switzerland.

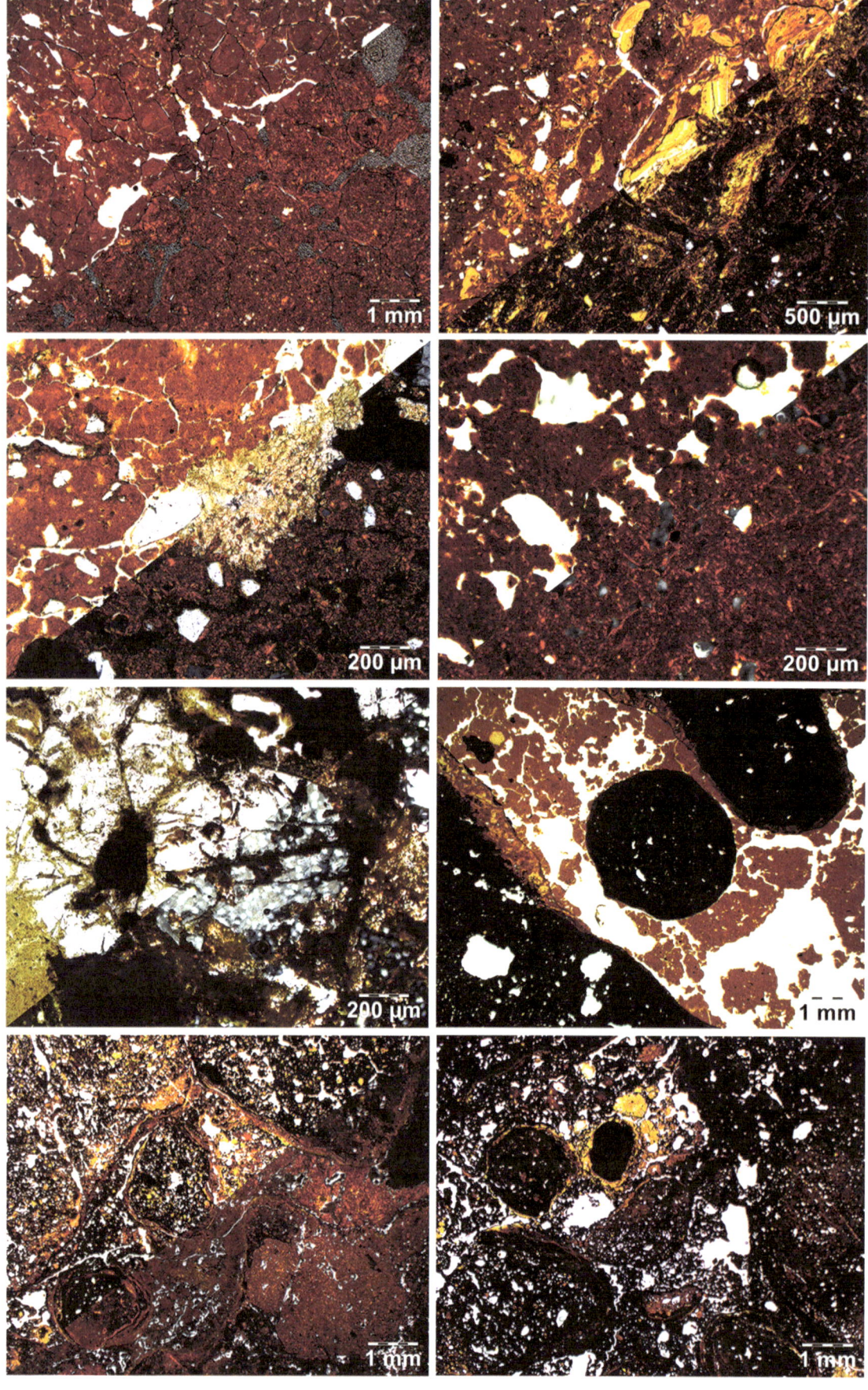

File 73: Processes Involving Iron Oxyhydroxides

Under intense geochemical weathering, oxyhydroxides can be redistributed inside the groundmass (Duchaufour 1998). This process usually turns the micromass red (rubification), which is a common trait of two distinct processes: fersiallitization and ferrallitization (Duchaufour 1977; Schaetzl and Thompson 2015). In addition, due to the strong weathering of primary minerals, clays can form and become the main compound of the micromass. An exhaustive description of mineral weathering at the microscale can be found in Delvigne (1998).

Captions from upper left corner to lower right corner.

1. Red homogeneous and clayey micromass due to the process of rubification in a fersiallitic soil. Note the low birefringence of the b-fabric. Rhodic Luvisol, Apulia, Italy.

2. Clay coatings due to clay translocation inside a fersiallitic soil. Clay translocation can also affect rubified soils subjected to high seasonal temperature and moisture variations. The clay accumulation leads to the formation of stress-deformed clay coatings due to argilliturbation (see "File 69"). Paleo-Chromic Luvisol, Lombardy, Italy.

3. During the dry season, calcium carbonate can precipitate in fersiallitic soils. Such carbonate nodules are often microsparitic. Paleo-Chromic Luvisol, Lombardy, Italy.

4. In the presence of kaolinite (a TO clay), round sand-sized micro-aggregates of oxyhydroxides and clays form, inducing a specific microstructure called "pseudo-sand" (Ahn 1970), which is typical of a ferrallitic soil. Acrisol, Mongodara district, Burkina Faso.

5. Weathering of a quartz grain forming a "runiquartz" (Eswaran et al. 1975). Parts of the quartz grain are infilled by oxyhydroxides (dark brown micromass). Paleosol, Ligurian coast, northern Italy.

6. Large nodules of iron oxyhydroxides (in black) surrounded by a clay micromass (light brown). This type of structure appears inside tropical soils, which are not yet completely encrusted by iron oxyhydroxides. PPL view, Plinthosol, Mongodara district, Burkina Faso.

7.–8. Typical characteristics of iron-encrusted tropical soils containing coalescent oxyhydroxide nodules and partially surrounded by clay coatings. In 8., only PPL view, showing the dense oxyhydroxide composition of some nodules and the groundmass. Plinthosol, Fort Portal district, Uganda.

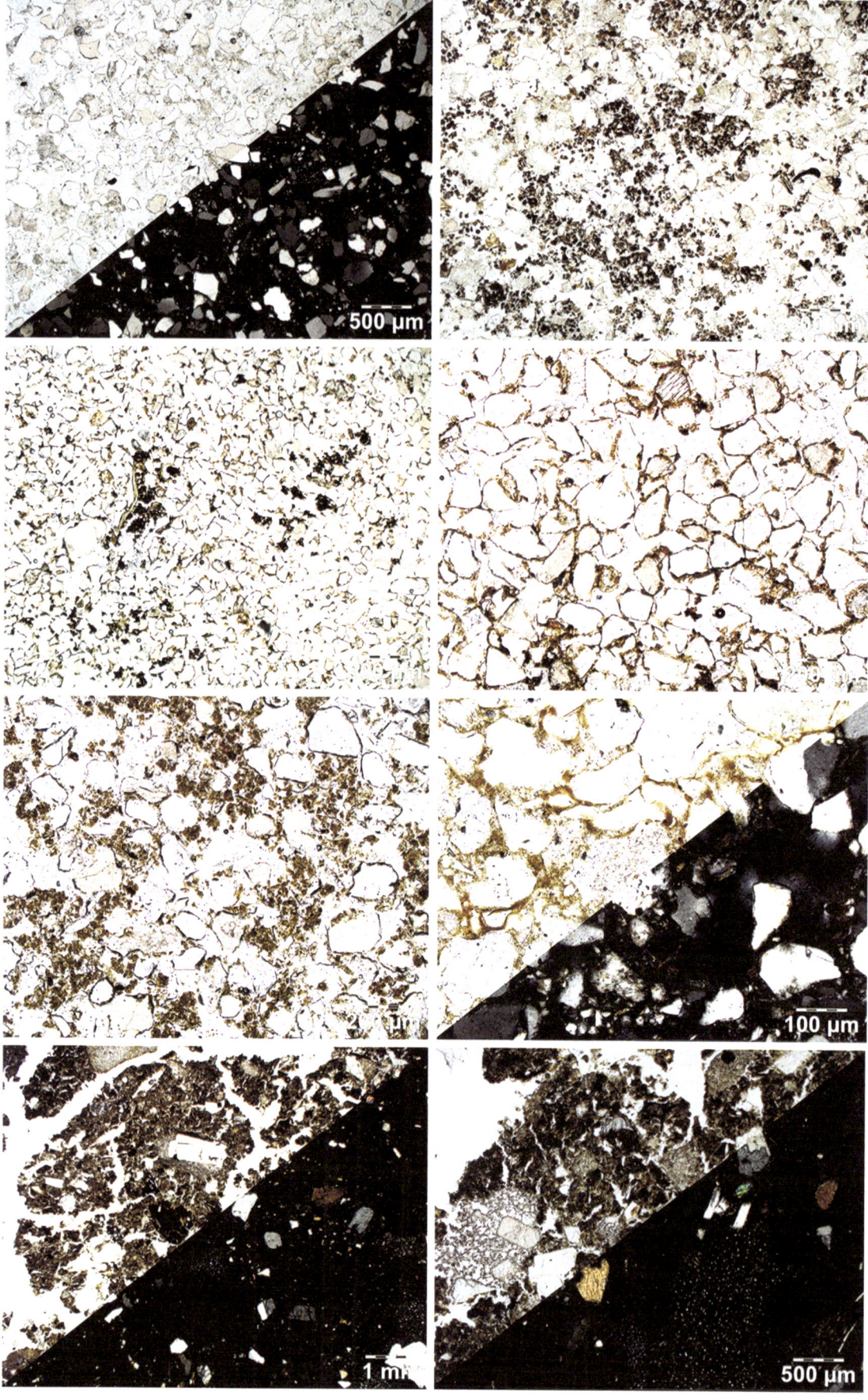

File 74: Biogeochemical Processes I

During biogeochemical processes, organic matter can combine with mineral compounds to form organomineral complexes that give soils specific properties. In addition, organic matter constitutes a driving force of pedogenesis as it influences the processes of weathering, as well as the transfer of matter inside profiles (Duchaufour 1997). In soil micromorphology, podzolization and andosolization are clearly expressed by specific features. In the case of podzols, horizons display striking micromorphological differences, while in andosols, micromorphological characteristics are found more in the general aspect of the thin section.

Captions from upper left corner to lower right corner.

1.   Eluvial E horizon made of loosely packed quartz grains in a coarse monic c/f related distribution. Note the absence of mineral fine material and organic constituents. Podzol, L'Isle-Adam forest, Paris Basin, France.

2.   Bh horizon showing a coarse groundmass made of quartz grains associated with an organomineral micromass organized in micro-aggregates in an enaulic c/f related distribution and referred to as polymorphic material (Van Ranst et al. 2018). Podzol, L'Isle-Adam forest, Paris Basin, France.

3.   Bhs horizon showing a coarse groundmass made of quartz grains associated with clusters of dark organomineral matter organized in micro-aggregates. Some brownish material starts to bridge quartz grains, making the c/f related distribution intermediate between enaulic and gefuric. PPL view, Podzol, L'Isle-Adam forest, Paris Basin, France.

4.   Bs horizon with quartz grains coated by yellowish to dark brown monomorphic Fe- and/or Al-rich material (Van Ranst et al. 2018), slightly cracked or uncracked. Coatings can include amorphous organic matter. The c/f related distribution is chitonic. PPL view, Podzol, L'Isle-Adam forest, Paris Basin, France.

5.   Bs horizon showing an enaulic c/f related distribution with polymorphic fine mineral material mixed with oxyhydroxides. Such soils are intergrades between Cambisols and Podzols (Duchaufour 1997). PPL view, Entic Podzol, L'Isle-Adam forest, Paris Basin, France.

6.   Same horizon as in 5. but characterized by a chitonic c/f related distribution with a very fine monomorphic mineral micromass, also forming coatings. Entic Podzol, L'Isle-Adam forest, Paris Basin, France.

7.–8.   In these thin sections, the micromass is totally extinct. Only some coarse mineral grains can be identified in XPL (e.g. amphiboles; see "File 33"). This extinction is due to the amorphous composition of the micromass, which is dominated by allophanes and other short-range order minerals associated to organic matter. Andosols, Chaîne des Volcans, Massif Central, France.

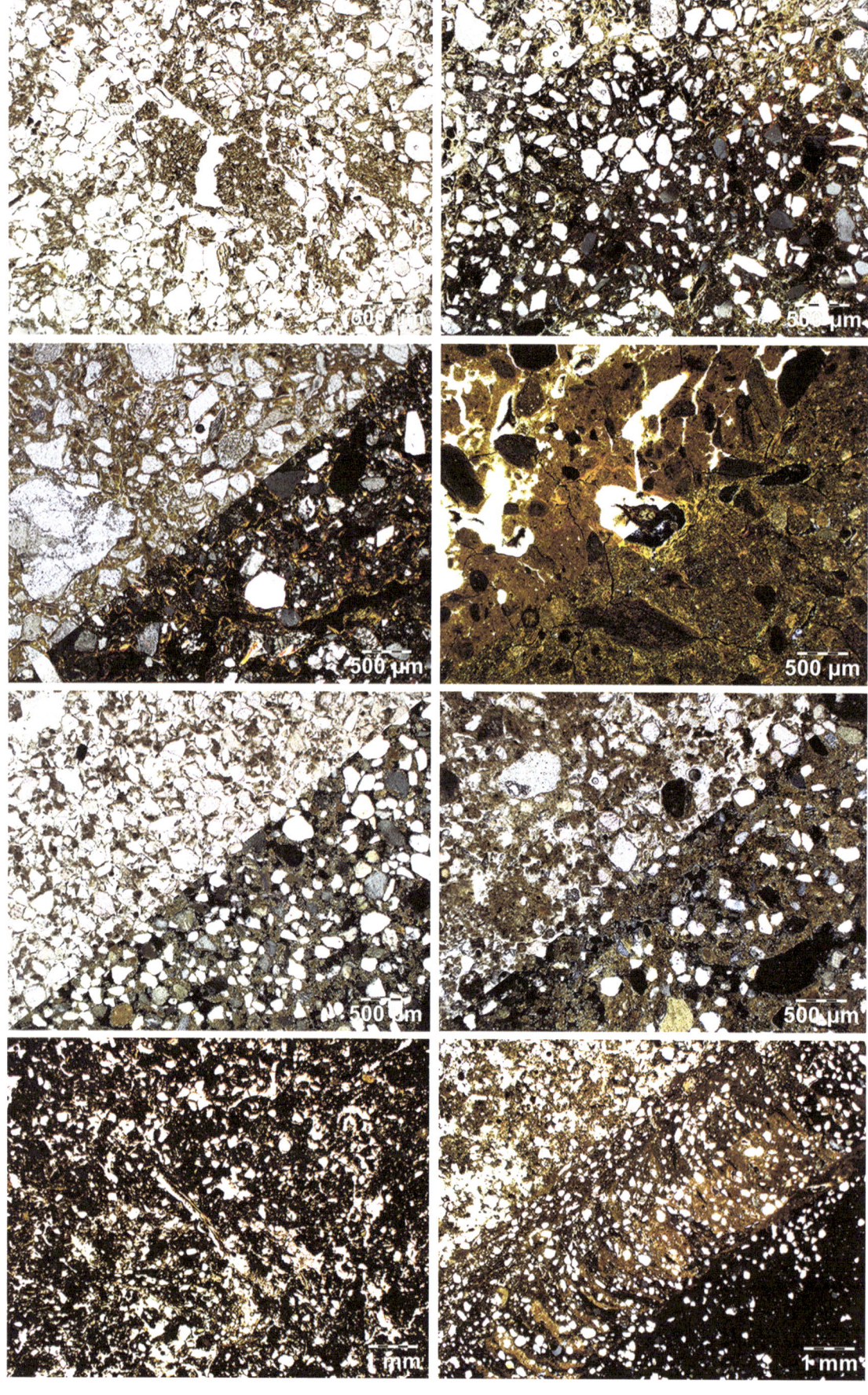

File 75: Biogeochemical Processes II

Brunification is a common biogeochemical process observed in temperate regions, during which clay minerals, iron, and organic matter interact to form clay–humic and clay–iron complexes. In environments with more pronounced contrasts between seasons, another type of biogeochemical process prevails, i.e. the deep incorporation of stable organic matter forming isohumic profiles. Unfortunately, soil micromorphology cannot provide any clear and univocal sets of diagnostic features for such kinds of processes. In this section, only a few examples of micromorphological aspects of soils under brunification and melanization are given.

Captions from upper left corner to lower right corner.

1. Quartz grains in a light brownish micromass composed of fine mineral material and organic matter, forming an irregularly-spaced porphyric c/f related distribution. PPL view, Dystric Cambisol, L'Isle-Adam forest, Paris Basin, France.
2. Quartz grains in a light to dark brownish micromass composed of fine mineral material and organic matter, forming a single-spaced porphyric c/f related distribution. In XPL, the birefringence pattern suggests an early phase of clay coating formation. Gleyic Cambisol, L'Isle-Adam forest, Paris Basin, France.
3. Coarse mineral grains in a clayey micromass forming a single-spaced porphyric c/f related distribution. The b-fabric is granostriated, and clay coatings are easily observed in XPL. Luvisol, Jura Mountains, Switzerland.
4. Brunification and clay translocation (see "File 69") can often be associated during soil evolution. Thin clay coatings formed in this brown groundmass made of clays, oxyhydroxides, and mixed with claystone fragments. They are easily identifiable by their specific birefringence. Paleo-Luvisol, Apennines, Italy.
5. Micro-aggregates formed by organic compounds and fine mineral matter. This porphyric c/f related distribution emphasizes the homogeneity in this thin section. Kastanozem, Chobe Enclave, Botswana.
6. Micro-aggregates of organic compounds mixed with fine minerals. Pellets and bio-aggregates are characteristic traits of this soil and are due to bioturbation by termites. Kastanozem, Chobe Enclave, Botswana.
7. Dark micromass composed of organic matter and fine material, forming a close porphyric c/f related distribution. The colour and the fabric are typical of a melanization process. PPL view, Chernozem, Chobe Enclave, Botswana.
8. Bow-like passage feature in a Chernozem, probably generated by termite bioturbation. This pedofeature (see "File 63") has a specific fabric, in which crescent-like layers are stacked. Chobe Enclave, Botswana.

Chapter 6

The Future of Soil Micromorphology

File 76: The Future of Soil Micromorphology

The advancement of technology opens up new opportunities to soil micromorphology. Although a description using an optical microscope of the fabric and the various constituents of soils will be always necessary to investigate soil evolution, the uncovered thin section leaves soil material on which analyses can be performed. Since the 1970s, it was possible to observe thin sections at high resolution with the scanning electron microscope in its backscattered electron mode (see "File 7"). It was also possible to generate chemical images with electron microprobes. But these conventional techniques, as well as new ones, greatly improve the study of matter interactions in soils, not only by enhancing the spatial resolution with incredible precision but also by providing chemical and mineralogical images, which substantially increased the accuracy of micromorphological diagnostics. By coupling morphological and chemical approaches, including stable isotope imaging in soil material, the future of soil micromorphology will undoubtedly offer new opportunities to solve specific problems, especially in the field of organomineral interactions in soils. It is wise to say that soil micromorphology, with its analytical and holistic approaches, will make it possible to build the necessary solid foundations needed for investigations that are increasingly oriented towards nanoscale objects: it will remind us that the trees should not hide the forest.

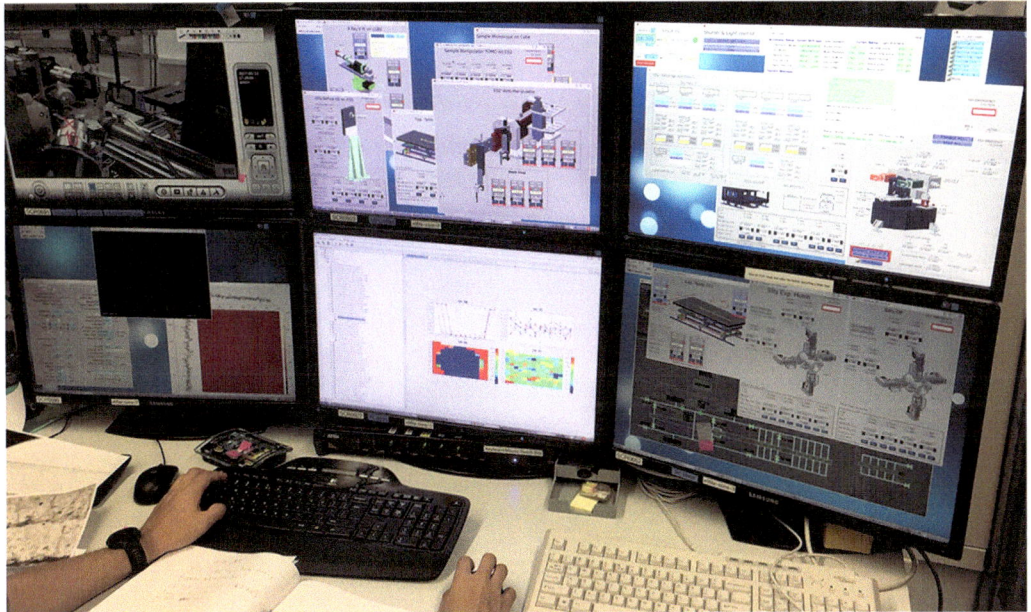

Cockpit of the microXAS Beamline at the Swiss Light Source, Paul Scherrer Institute, Villigen PSI, Switzerland.

© The Author(s) 2021
E. P. Verrecchia, L. Trombino, *A Visual Atlas for Soil Micromorphologists*,
https://doi.org/10.1007/978-3-030-67806-7_6

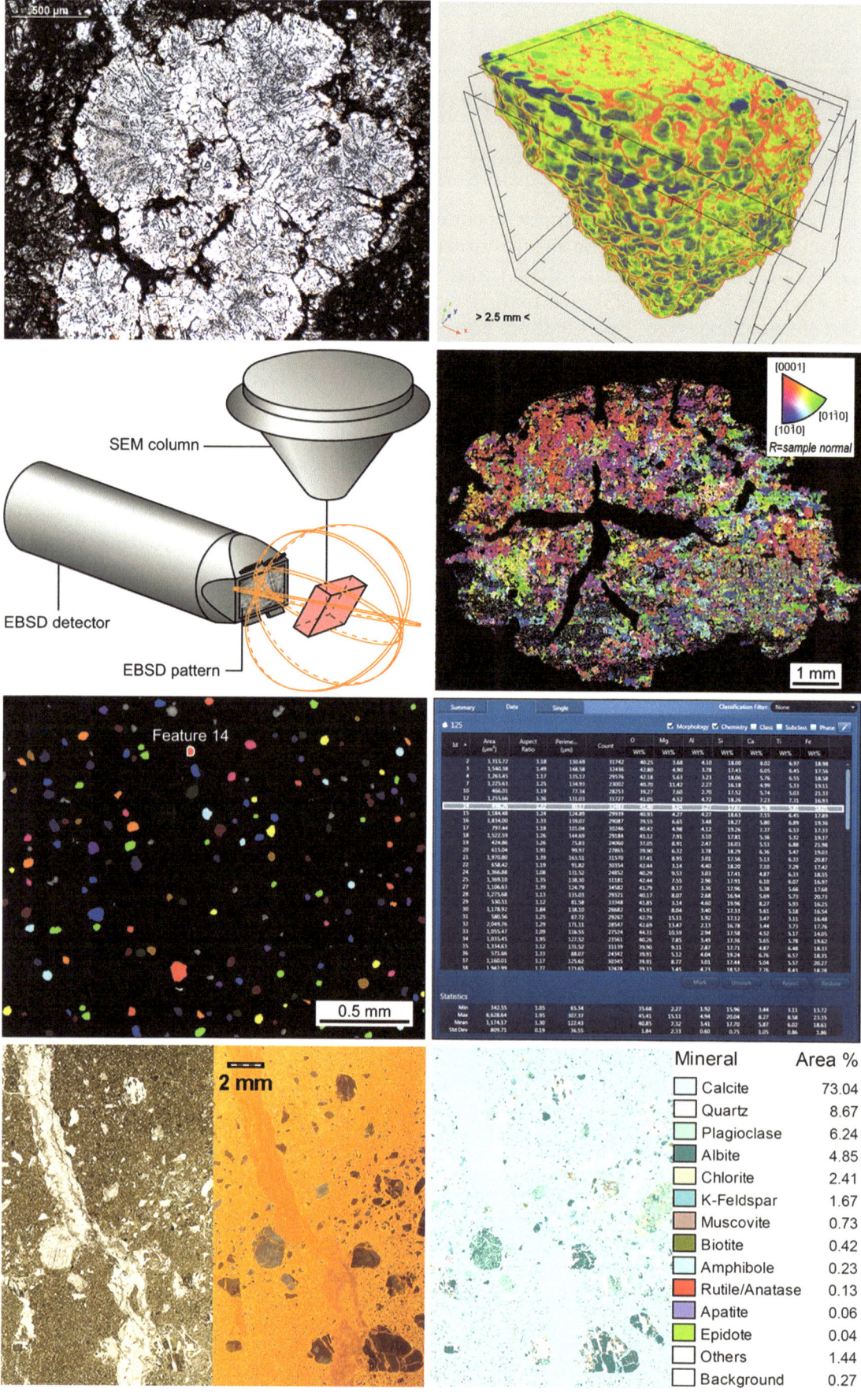

Mineral	Area %
Calcite	73.04
Quartz	8.67
Plagioclase	6.24
Albite	4.85
Chlorite	2.41
K-Feldspar	1.67
Muscovite	0.73
Biotite	0.42
Amphibole	0.23
Rutile/Anatase	0.13
Apatite	0.06
Epidote	0.04
Others	1.44
Background	0.27

File 77: **Beyond the Two Dimensions**

Based on density variation and X-ray attenuation, X-ray microtomography makes it possible to visualize the organization of voids and particles of different natures in a soil volume. Another tool, electron backscatter diffraction (EBSD), can be used to detect the crystallographic axes of minerals in a thin section of soil. Finally, a device such as the QEMSCAN opens the avenue of quantitative analysis of minerals in thin sections of soil.

Captions from upper left corner to lower right corner.

1. View in PPL of *Microcodium* in a paleosol from the Corbières, southern France.
2. Soil 3D-block at the same location as 1. observed with a high energy micro-CT scanner. This instrument produces a three-dimensional X-ray attenuation map for rock or soil samples in a non-destructive way. In this case, it is possible to see that *Microcodiums* form a dense and intertwined mesh of biominerals.
3. Sketch showing the generation of an EBSD Kikuchi pattern from the interaction between the SEM electron beam and the sample's crystal lattice. Within the sample, elastic and inelastic scattering produces a three-dimensional interaction volume of multidirectional electrons. When the Bragg conditions are satisfied, constructive interference occurs, generating straight and parallel line pairs when projected onto the EBSD detector screen. The association of bands forms an EBSD Kikuchi pattern from which the crystallographic orientation of the material can be determined. Courtesy of Dr. Pierre Vonlanthen (University of Lausanne).
4. EBSD inverse pole figure map of a calcite nodule (Cameroon) showing orientations of calcite crystallographic axes with respect to a reference sample. Here, pixels coloured in red are those for which calcite [0001] axes are (closely) parallel to the sample normal. The occurrence of nearby grains with similar colours indicates a non-random distribution of crystallographic orientations. For a detailed review of the EBSD technique, see Prior et al. (1999) or Schwarzer et al. (2009).
5. Automatic morphological and chemical analyses of soil particles performed in the SEM using the AZtec Feature software plug-in by Oxford Instruments. Particles are detected in the backscattered electron image based on greyscale levels and site-specifically analysed for chemistry using X-ray energy-dispersive spectroscopy.
6. Lookup table listing the morphological parameters and elemental composition corresponding to each individual particle.
7. PPL (left) and cathodoluminescence views of a carbonate-rich thin section.
8. QEMSCAN image of the same view as 7. showing the distribution of the various minerals detected and their relative proportions. Calcite dominates the mineral fraction of this thin section, followed by quartz and plagioclase. Courtesy of Dr. Kalin Kouzmanov (University of Geneva).

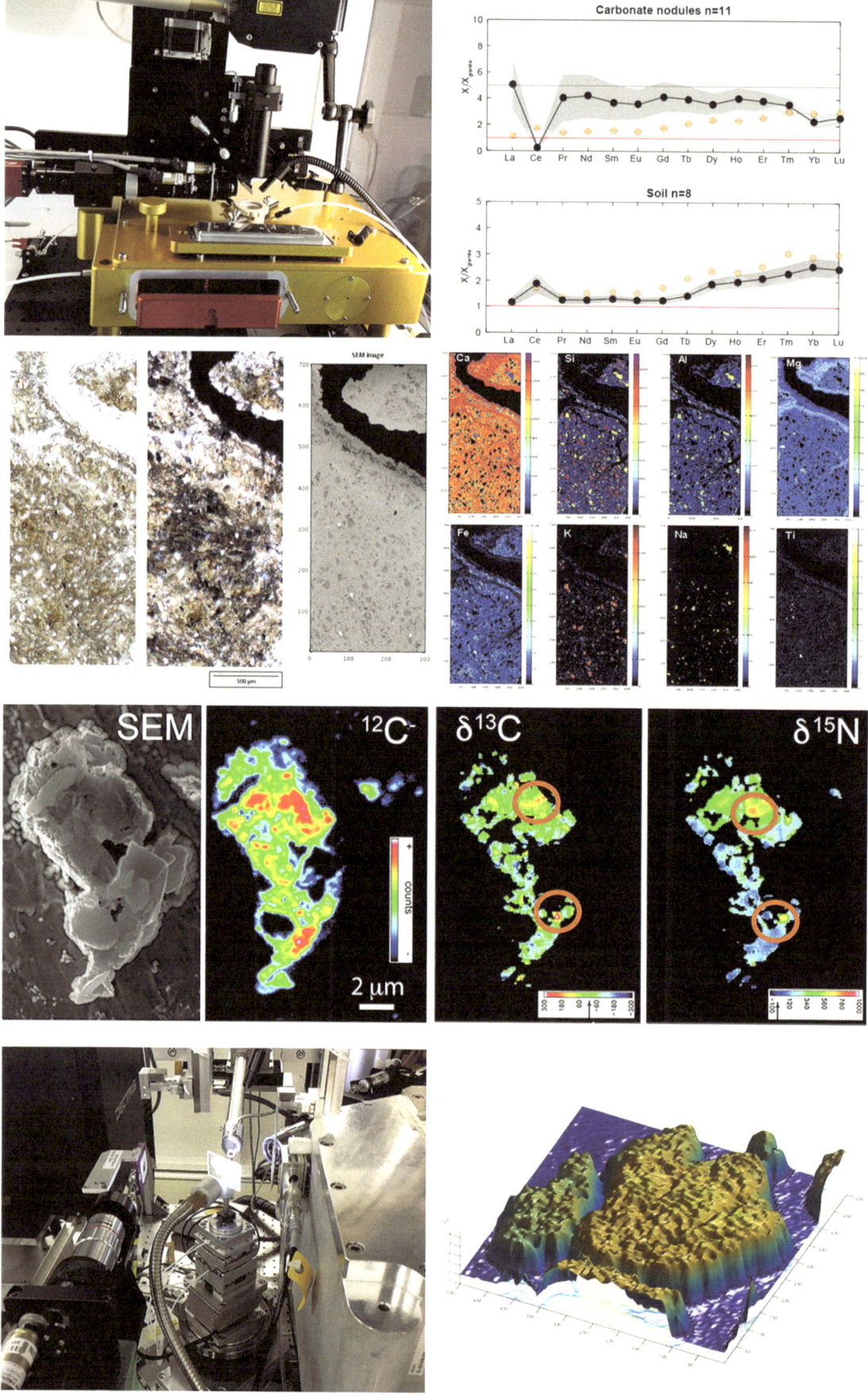

File 78: The Prospect of Chemical Imaging

It must not be forgotten that uncovered thin sections are still a natural soil sample. Minerals and organic matter remain accessible for measurements, and this is particularly true with modern instruments, with which it is possible to map the chemical and mineralogical compositions of the various components forming the soil. This section provides a few examples of chemical measurements made on sections or small soil objects, performed with different instruments.

Captions from upper left corner to lower right corner.

1. S155 two-volume ablation cell of a RESOlution ablation system. A sample shuttle for custom-sized samples (in red) is partly inserted in the cell (in yellow). The ablated aerosol is extracted through the extraction funnel in the central part of the cell and directed towards the torch of the ICP-MS, where the ablated particles are heated and vaporized, and the chemical elements contained in them, ionized. Courtesy of Dr. Alexey Ulyanov (University of Lausanne).

2. Comparison of REE average distributions in carbonate nodules ($n = 11$; upper curve in black) and soils ($n = 8$; lower curve in black) with respect to Saharan dust (in orange). Dark grey shaded area refers to (1σ) standard deviation. Paleo-Vertisols from northern Cameroon (Dietrich et al. 2017). The REE abundance has been directly measured by laser-ablation ICP-MS with the instrument shown partially in photograph 1.

3. PPL, XPL, and scanning electron microscope in backscattered electron mode views of a carbonate nodule from a Vertisol (Cameroon).

4. Distribution maps of elements (Ca, Si, Al, Mg, Fe, K, Na, and Ti) obtained using wavelength-dispersive X-ray spectroscopy of the same sample as in 3. Warmer colours refer to relatively higher contents. Courtesy of Dr. Nathalie Diaz (University of Lausanne).

5.–6. NanoSIMS images of glycine-derived ^{13}C and ^{15}N spots identified at the surface of soil aggregates, randomly isolated from soil density fractions, separated from a surface forest soil, and incubated for 8 h with uniformly $^{13}C/^{15}N$-labelled glycine (Hatton et al. 2015). The arrow represents the natural values for ^{13}C and ^{15}N isotope ratios. Courtesy of Dr. Laurent Remusat (Muséum National d'Histoire Naturelle, France).

7. Thin section (bright light in the centre) placed under the microXAS beamline at the Swiss Light Source, Paul Scherrer Institute, Villigen PSI, Switzerland.

8. Map of cryptomelane distribution (low content in blue, high content in yellow) inside a digitate manganese nodule observed in a soil developed on travertine, Morocco. The nodule is approximately 100 μm in width.

Appendix A

Chemical Formulae of Some Minerals Observed in Soils

Mineral	Family or chemical name	Formula
Actinolite	Amphibole	$Ca(Mg, Fe^{2+})_5(Si_8O_{22})(OH)_2$
Anhydrite	Calcium sulphate	$CaSO_4$
Apatite	Calcium phosphate	$Ca_5(PO_4)3(OH, Cl, F)$
Aragonite	Calcium carbonate	$CaCO_3$
Augite	Pyroxene	$(Ca, Fe_{2+}, Mg)Si_2O_6$
Barite	Barium sulphate	$BaSO_4$
Biotite	Mica	$K_2(Mg, Fe^{2+})_4(Si_6Al_2O_{20})(OH)_4$
Calcite	Calcium carbonate	$CaCO_3$
Chalcedony	Silica	SiO_2
Chlorite	Phyllosilicate	$(Fe, Mg, Al)_6(Si, Al)_4O_{10}(OH)_8$
Diopside	Pyroxene	$CaMgSi_2O_6$
Dolomite	Ca and Mg double carbonate	$(Ca, Mg)(CO_3)_2$
Enstatite	Pyroxene	$Mg_2Si_2O_6$
Epidote	Epidote	$Ca_2(Fe^{3+}, Al)Al_2(SiO_4)(Si_2O_7)O(OH)$
Fayalite	Olivine	$Fe_2^{2+}SiO_4$
Ferrosilite	Pyroxene	$FeSiO_3$
Forsterite	Olivine	$Mg_2^{2+}SiO_4$
Glauconite	Mica	$(Fe^{3+}, Al, Mg)_2(Si, Al)_4O_{10}(OH)_2$
Goethite	Iron oxide	$FeO(OH)$
Grossular	Garnet	$Ca_3Al_2(SiO_4)_3$
Gypsum	Calcium sulphate	$CaSO_4 \cdot 2H_2O$
Halite	Chloride	$NaCl$
Hematite	Iron oxide	Fe_2O_3
Hornblende	Amphibole	$Ca(Mg, Fe^{2+})_4Al(Si_7AlO_{22})(OH)_2$
Ilmenite	Iron and titanium oxide	$(Fe, Mg, Mn)TiO_3$
Jadeite	Pyroxene	$NaAlSi_2O_6$
Kaolinite	Phyllosilicate	$Al_2Si_2O_5(OH)_4$
Kenyaite	Hydrous sodium silicate	$NaSi_{11}O_{20.5}(OH)_4 \cdot 3H_2O$
Magadiite	Hydrous sodium silicate	$NaSi_7O_{13}(OH)_3 \cdot 3H_2O$
Magnetite	Iron oxide	Fe_3O_4
Malachite	Copper carbonate	$Cu_2CO_3(OH)_2$
Microcline	Feldspar	KSi_3AlO_8
Montmorillonite	Phyllosilicate	$(Na, Ca)_{0.33}(Al, Mg)_2(Si_4O_{10})(OH)_2 \cdot n.H_2O$
Muscovite	Mica	$K_2Al_4(Si_6Al_2O_{20})(OH)_4$
Olivine	Olivine	X_2SiO_4 with $X = [Fe^{2+}, Mg^{2+}]$
Orthoclase	Feldspar	KSi_3AlO_8
Palygorskite	Phyllosilicate	$(Mg, Al)_2Si_4O_{10}(OH) \cdot 4(H_2O)$
Plagioclases	Feldspar	$Ca_xNa_{1-x}Al_{1+x}Si_{3-x}O_8$ with $x > 0.1$
Pyrite	Sulphide	FeS_2
Quartz	Silica	SiO_2
Sanidine	Feldspar	$(K, Na)AlSi_3O_8$
Sepiolite	Phyllosilicate	$Mg_4Si_6O_{15}(OH)_2 \cdot 6(H_2O)$
Sericite	Mica	$KAl_2(AlSi_3O_{10})(OH)_2$
Serpentine	Phyllosilicate	$Mg_3Si_2O_5(OH)_4$
Siderite	Carbonate	$FeCO_3$
Vivianite	Iron phosphate	$Fe_3^{2+}(PO_4)_2 \cdot 8H_2O$
Zircon	Zirconium silicate	$ZrSiO_4$

E. P. Verrecchia, L. Trombino, *A Visual Atlas for Soil Micromorphologists*,
https://doi.org/10.1007/978-3-030-67806-7

Errors and Pitfalls I

During thin section fabrication, some artefacts can be inadvertently generated, forming "features" that must be differentiated from true natural traits of soils. This section and the following one give some examples (not an exhaustive list of possible errors and pitfalls) of artefacts of different origins.

Captions from upper left corner to lower right corner.

1.–2. Bubbles in the mounting medium are caused by the presence of air between the cover and the support glass of the thin section. They are perfectly round with a high relief and not in focus with the soil material. Left: bubbles in a void; right: bubble affecting the soil groundmass.

3. Filament of a textile fibre as an example of external pollution of a thin section. Note that the fibre is not in focus with the soil material. Be careful not to confuse such textile fibres with fungal filaments.

4. Example of a hair trapped between the soil material and the cover glass. This is not a soil feature.

5. Dirt is often an artefact included within the soil material during preparation of the sample. Note that the dirt is not in focus with the soil material.

6. Dirt can also fill small-sized voids and planes, simulating the presence of amorphous material infilling. These artefacts are often randomly scattered in different parts of the thin section, without any regularity. Therefore, comparing parts that are affected and not affected is essential in identifying such artefacts.

7. When a plaster-impregnated burlap is used for sample collection (see "File 3"), some artefacts can be generated if the thin section includes part of the plaster envelope. The upper part of the microphotograph shows a large laminar structure due to successive layers of plaster-impregnated burlap.

8. Artificial porosity generated by disruption of the sample during drying and/or hardening.

© The Author(s) 2021
E. P. Verrecchia, L. Trombino, *A Visual Atlas for Soil Micromorphologists*,
https://doi.org/10.1007/978-3-030-67806-7

Errors and Pitfalls II

During the thin section fabrication, some artefacts can be inadvertently generated, forming features that must be differentiated from true natural traits of soils. This section and the former one give some examples (not an exhaustive list of possible errors and pitfalls) of artefacts of different origins.

Captions from upper left corner to lower right corner.

1. Variations in the thin section thickness due to an uneven polishing of the thin section surface. The lower left part of the microphotograph is far too thin, causing the progressive disappearance of soil groundmass.
2. Variations in the thin section thickness due to an uneven polishing of the thin section surface. In addition to an excessive thinning of the soil material, some of the groundmass has been ground away (lower right side).
3. The thin section is too thick; quartz grains display yellow interference colours.
4. Blurred area due to an unidentified cause, although the other part around it is in focus. This type of artefact makes getting an accurate focus difficult.
5. Grains of abrasive powder can be incorporated into the soil material. The shape of grains is angular and fairly constant with a high relief.
6. Misty to cloudy and greyish areas inside the resin due to some possible chemical reactions between the resin and the polishing fluids.
7.–8. When a thin section has been stored for a very long time, it is possible that the resin can sometimes form crystals from the polymeric material used (depending on the type of glue, the dilution rate, etc.). These crystals, only visible in XPL, have different sizes and shapes, from very small clusters (on the left side microphotograph) to coarser dendritic clumps.

© The Author(s) 2021
E. P. Verrecchia, L. Trombino, *A Visual Atlas for Soil Micromorphologists*,
https://doi.org/10.1007/978-3-030-67806-7

slide reference   [   ]

Microstructure and Porosity Patterns

aggregates

type
size
orientation
distribution
pedality
accomdtion

voids

type
frequency
size
orientation
distribution

microstructure type   [   ]

other aggregates

definition
shape

size
frequency
nature
weathering
orientation
distribution

Basic Mineral Components

coarse material

nature
frequency
size
shape
roughness
weathering
orientation
distribution

fine material

size
color
nature
limpidity
b-fabric

Groundmass

c/f limit
c/f distribution
c/f ratio

Basic Organic Components

coarse material

nature
frequency
size
shape
roughness

weathering

orientation
distribution

Basic Org. Comp. (follows)

fine material

nature
frequency
size
shape
orientation
distribution

pigments

shape
color
shape
color
shape
color

Pedofeatures

key

group
type
subtype
nature
frequency
size
color
related
fragmented

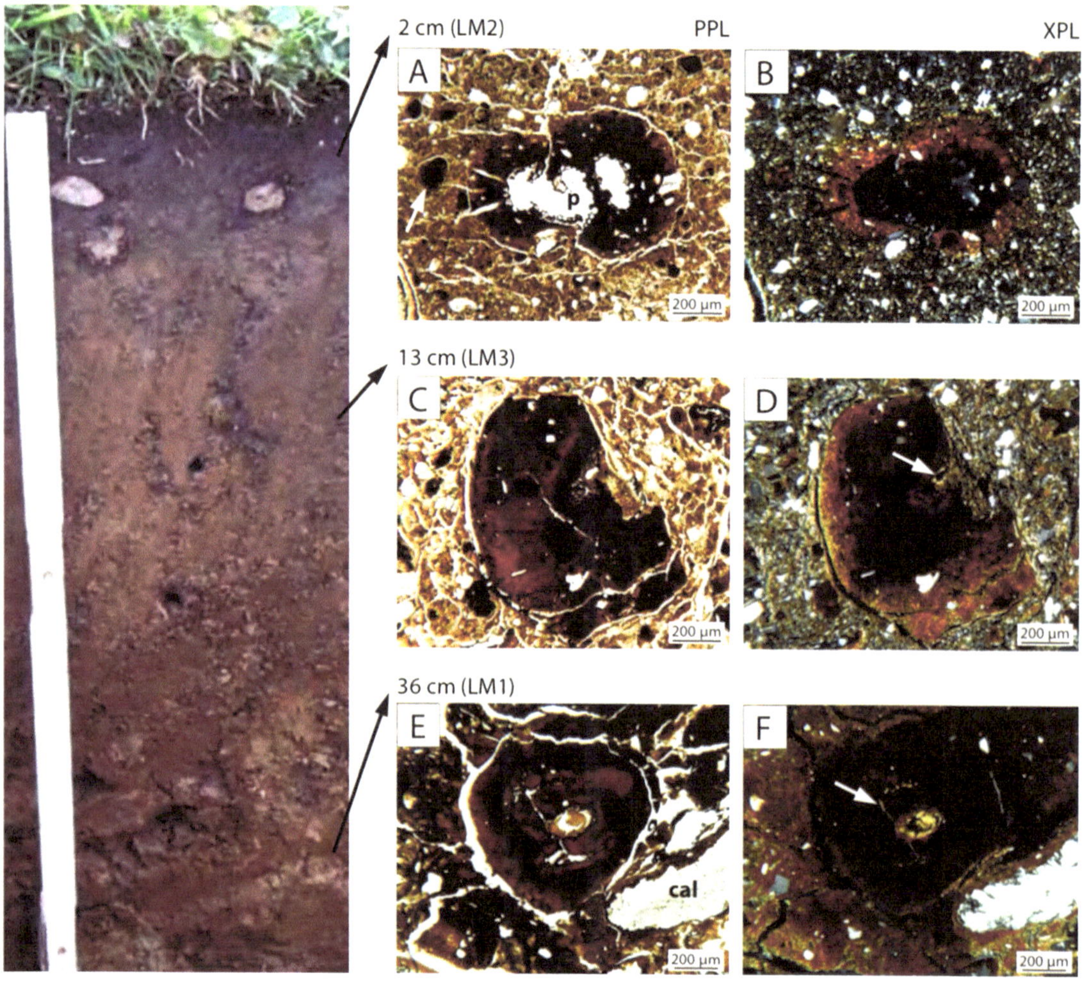

2 cm (LM2)   PPL   XPL

A   p   200 μm   B   200 μm

13 cm (LM3)

C   200 μm   D   200 μm

36 cm (LM1)

E   cal   200 μm   F   200 μm

How to Describe a Thin Section

> The description of a soil thin section can be extremely time consuming. Therefore, a good protocol will save a lot of time. This section introduces a succession of steps that can help to organize the thin section description and proposes two ways to comprehensively present the data for reports or publications. The first is in a table for listing the pertinent information, which can be processed with any spreadsheet software (Bullock et al. 1985); obviously, such a checklist needs to be re-created for each soil and site type. The second is a graphical summary of results, introduced by Kemp (1985). Both approaches are complementary and could be provided in a document.

The following steps can be used to make a preliminary and detailed description of a thin section (see also "File 9"):

1. First, it is wise to look at the thin section with the naked eye. Held up to the light, it is easy to identify up to four or five specific and apparently homogeneous areas. Also note the large objects (clearly different from the background), features, or traits. In addition, at the scale of the naked eye, sizes and shapes of aggregates are easily spotted. Afterwards, all these areas will be observed under the microscope, starting by using the lowest magnification.
2. Each large object with sharp boundaries should be observed and identified. They are usually coarse mineral materials or organic matter fragments.
3. For each more or less homogeneous area, identify its pattern and fabric, i.e. the voids, the aggregates, and the microstructures (see Chap. 2), and its main constituents, i.e. the coarse mineral and organic phases, the micromass (see Chap. 3), and its c/f related parameter (see "File 13" and "File 14").
4. The next step is the description of pedogenic features (see Chap. 4).
5. The last step consists of the interpretation of the identified parameters (see 1. below), objects, and features based on the observer's soil science background; in this Atlas, some examples of interpretations are given in Chap. 5.

Captions from top to bottom.

1. Example of a spreadsheet including different parameters in order to achieve the most complete, documented, and hierarchical description of the thin section. This comprehensive example of spreadsheet is available on the Atlas website.
2. Example of a composite picture showing the soil profile and the various and predominant features. The main characteristics of the micromorphological traits must be described in the figure captions using the vocabulary of Stoops (2003).

Multilingual List of Useful Micromorphological Terms

This multilingual lexicon provides the most useful terms that can be applied in soil micromorphology in four different languages, i.e. English, French, Italian, and German. The idea of a lexicon has been borrowed from Georges Stoops, who proposed one as early as 1986 (Stoops 1986). A new and up-to-date list was compiled in 2017 by Georges Stoops and different authors in 19 different languages (www.isric.online/explore/ISRIC-collections/micromulti). This list concentrates exclusively on the vocabulary used in this Atlas and, therefore, is not exhaustive compared to Stoops (2003, 2021). Some translations are not the same in the list available on the cited website and in this Atlas, due to some corrections.

English	French	Italian	German
Accommodation	Ajustement	Accomodamento	Passgenauigkeit
Aggregate	Agrégat	Aggregato	Aggregat
Alteration	Altération	Alterazione	Änderung
Alteromorph	Altéromorphe	Alteromorfo	Alteromorph
Angular	Anguleux	Angolare	Eckig
Anorthic	Anorthique	Anortico	Anorthisch
Arrangement	Assemblage	Disposizione (spaziale)	Anordnung
B-fabric	Trame de biréfringence	Fabric di birifrangenza	b-Gefüge
Banded	En bande	A bande	Bandförmige
Basic	De base	Fondamentale	Grund
Biospheroid	Biosphéroïde	Biosferoide	Bio-Sphäroid
Bistrial	Bistrié	Bistriale	Bigestreifter
Blocky	Polyédrique	Poliedrico	Polyeder
Bow-like	En arc de cercle	Ad arco	Bogenformige
Capping	En coiffe	Ricoprimento	Kappe
Chamber	Chambre	Camera	Kammer
Channel	Chenal	Canale	Gang
Chitonic	Chitonique	Chitonica	Chitonisch
Circular	Circulaire	Circolare	Kreisfö rmig
Close	Serrée	Chiusa	Enge
Cloudy	Nébuleux	Nebuloso	Trübe
Clustered	En grappes	A gruppi	Gehäufte
Coarse/fine (c/f)	Grossier/fin (g/f)	Grossolano/fine (g/f)	Gross/fein (g/f)
Coating	Revêtement	Rivestimento	Überzug
Complex	Complexe	Complesso	Komplexe
Compound	Composite	Composto	Zusammengesetzte
Concentric	Concentrique	Concentrico	Konzentrisch
Crescent	En croissant	A mezzaluna	Sichelförmige
Cross-striated	Striation entre-croisées	Striato incrociato	Kreuzstreifiges
Crumb	Grumeau	Grumo	Krümel
Crust	Croéte	Crosta	Kruste
Cryptocrystalline	Cryptocristallin	Criptocristallino	Kryptokristallin

Crystal intergrowth	Cristal intercalaire	Accrescimento cristallino	Kristalleinwachsung
Crystallitic	Cristallitique	Cristallitico	Kristallitisches
Deformed	Déformé	Deformato	Deformierte
Dendritic	Dentritique	Dendritico	Dendritisches
Dense	Dense	Denso	Dicht
Depletion	Déplétion	Di svuotamento	Verarmung
Disorthic	Disorthique	Disortico	Disorthisch
Dissolved	Dissous	Dissolto	Aufgelöste
Distribution	Distribution	Distribuzione	Verteilung
Dotted	Piqueté	Puntinato	Punktiert
Double-spaced	Doublement es-pacé	A spaziatura doppia	Doppelabständige
Enaulic	Enaulique	Enaulica	Enaulisch
Fabric	Assemblage	Organizzazione Spaziale	Bodengefüge
Fan-like	En éventail	A ventaglio	Fächerförmige
Fragmented	Fragmenté	Frammentato	Fragmentiertes
Gefuric	Géfurique	Gefurica	Gefurisch
Geodic	Géodique	Geodico	Geodisch
Granostriated	Granostrié	Granostriato	Kornstreifiges
Granule	Granule	Granulo	Körnchen
Groundmass	Masse basale	Massa di fondo	Grundmasse
Hypo-coating	Hypo-revêtement	Iporivestimento	Hypo-Belag
Imbricated	Imbriqué	Embricato	Dachziegelartige
Impregnative	D'imprégnation	Di impregnazione	Imprägnierung
Inclined	Oblique	Inclinato	Schiefe
Infilling	Remplissage	Riempimento	Füllung
Intercalation	Intercalation	Intercalazione	Einschaltung
Interlaced	Entrelacé	Intrecciato	Netzartige
Intrusive	Intrusif	Intrusivo	Intrusion
Juxtaposed	Juxtaposé	Giustapposto	Nebeneinander
Laminated	Lamellaire	Laminato	Laminier
Layered	Lité	Stratificato	Geschichtet
Lenticular	Lenticulaire	Lenticolare	Lentikular
Limpid	Limpide	Limpido	Klar
Limpidity	Limpidité	Limpidezza	Klarheit
Linear	Linéaire	Lineare	Lineare
Lithomorphic	Lithomorphe	Litomorfico	Lithomorph
Loose	Lâche	Sciolto	Locker
Macrocrystalline	Macrocristallin	Macrocristallino	Makrokristallin
Mamillated	Mamelonné	Mammellonato	Traubige
Massive	Massive	Massivo	Massiv
Matrix	Matrice	Matrice	Matrix
Microcrystalline	Microcristalline	Microcristallino	Microkristallin
Microlaminated	Micro-lamellaire	Micolaminato	Mikrolaminiert
Micromass	Micromasse	Micromassa	Feinmasse
Micropan	Croéte interne	Crosta interna	Mikrokruste
Microstructure	Microstructure	Microstruttura	Mikrogefüge
Monic	Monique	Monica	Monisch

Monostriated	Monostrié	Monostriata	Einstreifiges
Mosaic speckled	Piqueté en mosaïque	Maculata a mosaico	Mosaikartig geflecktes
Nodule	Nodule	Nodulo	Nodul
Nonlaminated	Non-lamellaire	Non laminato	Unlaminiert
Nucleic	A nucleus	Nucleico	Mit Kern
Opaque	Opaque	Opaco	Undurchsichtigen
Open	Ouvert	Aperto	Offene
Organ residue	Résidu organique	Residuo di organo	Organrest
Orientation	Orientation	Orientazione	Orientierung
Orthic	Orthique	Ortico	Orthisch
Packing	Entassement	Intergranulare	Packung
Parallel	Parallèle	Parallelo	Parallele
Passage feature	Trait de passage	Figura di passaggio	Passagemerkmal
Pattern	Modèle	Modello	Muster
Ped	Agrégat	Aggregato	Aggregat
Pedofeature	Trait pédologique	Figura pedologica	Pedofeature
Pedomorphic	Pédomorphe	Pedomorfico	Pedomorph
Pendent	Pendeloque	Pendente	Hängebelag
Perpendicular	Perpendiculaire	Perpendicolare	Senkrechte
Plane	Fente	Vuoto planare	Riss
Plate	Lamellaire	Lamina	Platte
Pore	Vide	Poro	Hohlraum
Porostriated	Porostrié	Porostriato	Porenstreifiges
Porphyric	Porphyrique	Porfirica	Porphyrisch
Prism	Prisme	Prisma	Prisma
Quasicoating	Quasi-revêtement	Quasi-rivestimento	Quasi-Belag
Radial	Radial	Radiale	Radiale
Random	Aléatoire	Casuale	Zufallsverteilung
Random striated	A stries aléatoires	Striato disordinato	Zufällig gestreiftes
Separated	Séparé	Separato	Getrennt
Septaric	Septaria	Septarico	Septarisch
Serrated	Crénelé	Dentellato	Gezähnt
Simple	Simple	Semplice	Einfacher
Single grain	Entassement de grains	A grani singoli	Einzelkorn
Single-spaced	Espacement simple	A spaziatura singola	Einfach-abständige
Speckled	Tacheté	Maculato	Gefleckt
Spheroidal	Sphéroïdale	Sferoidale	Konzentrisch
Spongy	Spongieux	Spugnoso	Schwammik
Stipple speckled	En taches isolées	Maculato a puntini	Getüpfelt
Strial	Strié	Striale	Durchgängig
Striated	Strié	Striato	Streifige
Subangular	Subangulaire	Subangolare	Winfelteilt
Superimposed	Superposé	Sovrimposto	Überlagernde

Tissue	Tissu	Tessuto	Gewebe
Typic	Typique	Tipico	Typisch
Undifferentiated	Indifférencié	Indifferenziato	Undifferenziertes
Unistrial	Monostrié	Unistriale	Kreisförmig-streifiges
Vermicular	Vermiculaire	Vermicolare	Vermikular
Vesicle	Vésicule	Vescicola	Vesikel
Vesicular	Vésiculaire	Vescicolare	Vesikel
Void	Vide	Vuoto	Hohlraum
Vugh	Cavité	Vacuo	Kaverne
Vughy	Cavitaire	A vacui	Kavernen

References

Aassoumi, H., Broutin, J., Wartiti, M. E., Freytet, P., Koeniguer, J., Quesada, C., et al. (1992). Pedological nodules with cone in cone structure in the Permian of Sierra Morena (Spain) and Central Morocco. *Carbonates and Evaporites, 7*(2), 140–149.

Ahn, P. (1970). *West African soils*. Oxford: Oxford University Press.

Baran, E., & Monje, P. (2008). Oxalate biominerals. In *Biomineralization: From nature to application* (Vol. 4). Chichester: Wiley.

Becze-Deak, J., Langhor, R., & Verrecchia, E. (1997). Small-scale secondary $CaCO_3$ accumulations in selected sections of the European loess belt. Morphological forms and potential for paleoenvironmental reconstruction. *Geoderma, 76*, 221–252.

Benyarku, C., & Stoops, G. (2005). Guidelines for preparation of rock and soil thin sections and polished sections. Quaderns Dmacs 33, Departament de Medi Ambient i Ciencies del Sol, Univertat de Lleida, Spain

Bindschedler, S., Cailleau, G., & Verrecchia, E. (2016). Role of fungi in the biomineralization of calcite. *Minerals, 6*, 1–19

Bindschedler, S., Millière, L., Cailleau, G., Job, D., & Verrecchia, E. (2012). An ultrastructural approach to analogies between fungal structures and needle fiber calcite. *Geomicrobiology Journal, 29*(4), 301–313

Boggs, S., & Krinsley, D. (2006). *Application of cathodoluminescence imaging to the study of sedimentary rocks*. New York: Cambridge University Press.

Brantley, S., Goldhaber, M., & Ragnarsdottir, K. (2007). Crossing disciplines and scales to understand the critical zone. *Elements, 3*, 307–314.

Brewer, R. (1964). *Fabric and mineral analysis of soils*. London: Wiley.

Bullock, P., Fedoroff, N., Jongerius, A., Stoops, G., & Tursina, T. (1985). *Handbook for soil thin section description*. Wolverhampton: Waine Research Publications.

Cailleau, G., Braissant, O., Dupraz, C., Aragno, M., & Verrecchia, E. (2005). Biologically induced accumulations of $CaCO_3$ in Orthox soils of Biga, Ivory Coast. *Catena, 59*, 1–17.

Canti, M. (1998). The micromorphological identification of faecal spherulites from archaeological and modern materials. *Journal of Archaeological Science, 25*, 435–444.

Chesworth, W. (Ed.). (2008). *Fibric, hemic and sapric* (pp. 270–270). Dordrecht: Springer.

Delvigne, J. (1998). *Atlas of micromorphology of mineral alteration and weathering*. Number 3 in Spec. Publ. Ottawa: Mineralogical Assoc. of Canada and Paris: ORSTOM.

Dietrich, F., Diaz, N., Deschamps, P., Ngatcha, B. N., Sebag, D., & Verrecchia, E. P. (2017). Origin of calcium in pedogenic carbonate nodules from silicate watersheds in the Far North Region of Cameroon: Respective contribution of in situ weathering source and dust input. *Chemical Geology, 460*, 54–69.

Douglas, L., & Thompson, M. (1985). *Soil micromorphology and soil classification*. Number 15 in SSSA Special Publication. Madison, WI: Soil Science Society of America.

Duchaufour, P. (1977). *Pédologie 1*. Paris: Masson.

Duchaufour, P. (1997). *Abrégé de Pédologie* (5eme ed.). Paris: Masson.

Duchaufour, P. (1998). *Handbook of pedology*. Boca Raton: CRC Press.

© The Author(s) 2021
E. P. Verrecchia, L. Trombino, *A Visual Atlas for Soil Micromorphologists*,
https://doi.org/10.1007/978-3-030-67806-7

Durand, N., Monger, H., Canti, M., & Verrecchia, E. (2018). Calcium carbonate features. In *Interpretation of micromorphological features of soils and regoliths* (2nd ed.). Amsterdam: Elsevier.

Eswaran, H., Sys, C., & Sousa, E. (1975). Plasma infusions-a pedological process of significance in the humid tropics. *Anales de Edafologia y Agrobiologia, 34*, 665–674.

Fedoroff, N. (1971). The usefulness of micropedology in paleopedology. In D. Yaalon (Ed.), *Paleopedology: Origin, nature and dating of paleosols*. Jerusalem. International Society of Soil Sciences and Israel Universities Press.

Fedoroff, N., & Courty, M.-A. (2012). Textural features and microfacies expressing temporary and permanent soil water saturation. In R. Poch-Claret, M. Casamitjana, & M. Francis (Eds.), *Proceedings of the 14th Intern. Working Meet. on Soil Micromorphology* (p. 1.1.K). Session I. Lleida: Editions i Publications de la Universitat de Lleida.

FitzPatrick, E. (1984). *Micromorphology of soils*. London: Chapman and Hall.

FitzPatrick, E. (1993). *Soil microscopy and micromorphology*. Chichester: Wiley.

Freytet, P., Aassoumi, H., Broutin, J., Wartit, M. E., & Toutin-Morin, N. (1992). Présence de nodules pédologiques à structure cone-in-cone dans le Permien continental du Maroc, d'Espagne méridionale et de Provence. Attribution possible à une activité bactérienne associée à des racines de Cordaites. *Comptes rendus de l'Académie des Sciences. Sér. 2, 315*(6), 765–771.

Freytet, P., & Verrecchia, E. (1998). Freshwater organisms that build stromatolites: A synopsis of biocrystallization by prokaryotic and eukaryotic algae. *Sedimentology, 45*(3), 535–563.

Goldberg, G., & Macphail, R. (2003). Short contribution: Strategies and techniques in collecting micromorphology samples. *Geoarchaeology, 18*(5), 571–578.

Hasinger, O., Spangenberg, J., Millière, L., Bindschedler, S., Cailleau, G., & Verrecchia, E. (2015). Carbon dioxide in scree slope deposits: A pathway from atmosphere to pedogenic carbonate. *Geoderma, 247–248*, 129–139.

Hatton, P.-J., Remusat, L., Zeller, B., Brewer, E. A., & Derrien, D. (2015). NanoSIMS investigation of glycine-derived C and N retention with soil organo-mineral associations. *Biogeochemistry, 125*(3), 303–313.

Heilbronner, R., & Barrett, S. (2014). *Image analysis in earth sciences*. Berlin: Springer.

IUSS and Working-Group-WRB. (2014). World reference base for soil resources 2014 - international soil classification system for naming soils and creating legends for soil maps. World Soil Resources Reports 106, FAO, Rome, Italy.

Jaillard, B., Guyon, A., & Maurin, A. (1991). Structure and composition of calcified roots, and their identification in calcareous soils. *Geoderma, 50*(3), 197–210.

Kemp, R. (1985). *Soil micromorphology and the quaternary*. Cambridge: Quaternary Research Association. Technical Guide 2.

Krumbein, W. C. (1941). Measurement and geological significance of shape and roundness of sedimentary particles. *Journal of Sedimentary Research, 11*, 64–72.

Kubiëna, W. (1938). *Micropedology*. Ames, IA: Collegiate Press.

Legros, J.-P. (2012). *Major soil groups of the world*. Boca Raton: CRC Press.

MacKenzie, W., Adams, A., & Brodie, K. (2017). *Rocks and minerals in thin section: A colour Atlas*. Boca Raton: CRC Press.

Mann, S. (2001). *Biomineralization*. Oxford: Oxford University Press.

Marshall, D. (1988). *Cathodoluminescence of geological materials*. Boston: Unwin Hyman.

Matthews, W., & Boyer, R. (1976). *Dictionary of geological terms*. New York: Anchor Press.

Murphy, C. (1986). *Thin section preparation of soils and sediments*. Berkhamsted: AB Academic Publishers.

Nicosia, C., & Stoops, G. (2017). *Archaeological soil and sediment micromorphology*. Chichester: Wiley Blackwell.

Pagel, M., Barbin, V., Blanc, P., & Ohnenstatter, D. (2000). *Cathodoluminescence in geosciences*. Berlin: Springer.

Pentecost, A. (2010). *Travertine*. Berlin: Springer.

Powers, M. (1953). A new roundness scale for sedimentary particles. *Journal of Sedimentary Petrology, 23*, 117–119.

Prior, D. J., Boyle, A. P., Brenker, F., Cheadle, M. C., Day, A., Lopez, G., et al. (1999). The application of electron backscatter diffraction and orientation contrast imaging in the SEM to textural problems in rocks. *American Mineralogist, 84*(11–12), 1741–1759.

Rabenhorst, M., & Wilding, L. (1986). Pedogenesis on the Edwards Plateau, Texas: III. New model for the formation of petrocalcic horizons. *Soil Science Society of America Journal, 50*(3), 693–699.

Richter, D., Götte, T., Götze, J., & Neuser, R. (2003). Progress in application of cathodoluminescence (CL) in sedimentary petrology. *Mineralogy and Petrology, 79*(3), 127–166.

Richter, deB, D., & Yaalon, D. (2012). The changing model of soil, revisited. *Soil Science Society of America Journal, 76*(3), 766–778.

Rowley, M., Grand, S., & Verrecchia, E. (2018). Calcium-mediated stabilisation of soil organic carbon. *Biogeochemistry, 137*, 27–49.

Schaetzl, R., & Thompson, M. (2015). *Soils, genesis and geomorphology* (2nd ed.). Cambridge: Cambridge University Press.

Schwarzer, R. A., Field, D. P., Adams, B. L., Kumar, M., & Schwartz, A. J. (2009). Present state of electron backscatter diffraction and prospective developments. In A. J. Schwartz, M. Kumar, B. L. Adams, & D. P. Field (Eds.), *Electron backscatter diffraction in materials science* (pp. 1–20). New York: Springer US.

Sebag, D., Verrecchia, E., Lee, S., & Durand, A. (2001). The natural hydrous sodium silicates from the northern bank of Lake Chad: Occurrence, petrology and genesis. *Sedimentary Geology, 139*, 15–31.

Skinner, H., & Ehrlich, H. (2017). Biomineralization. In *Treatise in geochemistry - Biogeochemistry* (2nd ed., Vol. 10). Amsterdam: Elsevier.

Stoops, G. (1986). Multilingual translation of the terminology used in the "handbook for soil thin section description". *Pedologie, 36*(3), 337–348.

Stoops, G. (2003). *Thin section preparation of soils and sediments*. Madison: Soil Science Society of America, Inc.

Stoops, G. (2021). *Guidelines for analysis and description of soil and regolith thin sections*. Second Edition, Wiley.

Stoops, G., & Jongerius, A. (1975). Proposal for a micromorphological classification of soil materials. I. A classification of related distribution of fine and coarse particles. *Geoderma, 13*(3), 189–199.

Stoops, G., Marcelino, V., & Mees, F. (Eds.). (2018). *Interpretation of micromorphological features of soils and regoliths* (2nd ed.). Amsterdam: Elsevier

Van Ranst, E., Wilson, M., & Righi, D. (2018). Spodic materials. In *Interpretation of micromorphological features of soils and regoliths* (2nd ed.). Amsterdam: Elsevier.

van Vliet-Lanoë, B., & Fox, C. (2018). Frost action. In *Interpretation of micromorphological features of soils and regoliths* (2nd ed.). Amsterdam: Elsevier.

Verrecchia, E., Dumont, J., & Verrecchia, K. (1993). Role of calcium oxalate biomineralization of fungi in the formation of calcrete: A case study from Nazareth, Israel. *Journal of Sedimentary Petrology, 63*(5), 1000–1006.

Verrecchia, E., Freytet, P., Verrecchia, K., & Dumont, J. (1995). Spherulites in calcrete laminar crusts: Biogenic $CaCO_3$ precipitation as a major contributor to crust formation. *Journal of Sedimentary Research, A65*(4), 690–700.

Villagran, X., Mentzer, D. H. S., Miller, C., & Jans, M. (2017). Bone and other skeletal tissues. In C. Nicosia & G. Stoops (Eds.), *Archaeological soil and sediment micromorphology*. Chichester: Wiley Blackwell.

Villagran, X., & Poch, R. (2014). A new form of needle-fiber calcite produced by physical weathering of shells. *Geoderma, 2013*, 173–177.

Wadell (1932). Volume, shape, and roundness of rock particles. *Journal of Geology, 40*, 443–451.

Weil, R., & Brady, N. (2017). *The nature and properties of soils*. Harlow: Pearson Education Ltd.

Wieder, M., & Yaalon, D. (1974). Effect of matrix composition on carbonate nodule crystallisation. *Geoderma, 11*, 95–121.

Index

© The Author(s) 2021
E. P. Verrecchia, L. Trombino, *A Visual Atlas for Soil Micromorphologists*,
https://doi.org/10.1007/978-3-030-67806-7